Yeast Biotechnology

Special Issue Editor
Ronnie G. Willaert

MDPI • Basel • Beijing • Wuhan • Barcelona • Belgrade

MDPI

Special Issue Editor
Ronnie G. Willaert
Vrije Universiteit Brussel
Belgium

Editorial Office
MDPI AG
St. Alban-Anlage 66
Basel, Switzerland

This edition is a reprint of the Special Issue published online in the open access journal *Fermentation* (ISSN 2311-5637) from 2015–2017 (available at: http://www.mdpi.com/journal/fermentation/special_issues/saccharomyce).

For citation purposes, cite each article independently as indicated on the article page online and as indicated below:

Author 1; Author 2. Article title. *Journal Name* **Year**, *Article number*, page range.

First Edition 2017

ISBN 978-3-03842-442-0 (Pbk)
ISBN 978-3-03842-443-7 (PDF)

Photo courtesy of Dr. Ronnie G. Willaert

Table of Contents

About the Special Issue Editor

Ronnie Willaert, Dr. ir., Research Professor, has an extensive expertise in yeast research (*Saccharomyces cerevisiae, S. pastorianus, Candida albicans and C. glabrata*) and single-molecule biophysics (high-resolution microscopy, i.e., confocal laser microscopy, AFM, force spectroscopy, and scanning probe lithography), yeast space biology research and hardware development, protein science (yeast adhesins), cell (yeast) immobilisation biotechnology, fermentation technology, brewing science and technology, and mathematical analysis.

fermentation   **MDPI**

Editorial

Yeast Biotechnology

Ronnie G. Willaert [1,2,3,*]

1   Research Group Structural Biology Brussels, Vrije Universiteit Brussel, Brussels 1050, Belgium
2   Alliance Research Group NanoMicrobiology (NAMI), Vrije Universiteit Brussel, Brussels 1050,
    Belgium—Ghent University, Ghent 9000, Belgium
3   International Joint Research Group BioNanotechnology & NanoMedicine (NANO), Vrije Universiteit
    Brussel—Ecole Polytechnique de Lausanne (EPFL), Brussels 1050, Belgium—Lausanne 1015, Switzerland
*   Correspondence: Ronnie.Willaert@vub.ac.be; Tel.: +32-2-629-1846

Academic Editor: Badal C. Saha
Received: 13 January 2017; Accepted: 24 January 2017; Published: 26 January 2017

Keywords: *Saccharomyces cerevisiae*; non-*Saccharomyces* yeasts; fermentation-derived products; wine;
beer; flavour; bioethanol; enzyme production; bioreactors; nanobiotechnology

---

Yeasts are truly fascinating microorganisms. Due to their diverse and dynamic activities, they have been used for the production of many interesting products, such as beer, wine, bread, biofuels, and biopharmaceuticals. *Saccharomyces cerevisiae* (brewers' or bakers' yeast) is the yeast species that is surely the most exploited by man. *Saccharomyces* is a top choice organism for industrial applications, although its use for producing beer dates back to at least the 6th millennium BC. Bakers' yeast has been a cornerstone of modern biotechnology, enabling the development of efficient production processes. Today, diverse yeast species are explored for industrial applications. This Special Issue is focused on some recent developments of yeast biotechnology, i.e., bioethanol, wine and beer, and enzyme production. Additionally, the new field of yeast nanobiotechnology is introduced and reviewed.

New developments in efficient bio-ethanol production. Due to its low costs and wide distribution, lignocellulosic biomass is the most promising feedstock to be used in biorefineries and lignocellulose-derived fuels. The fermentation of sugars from lignocelluloses has been proposed as a viable pathway for the production of renewable biofuels. However, the feedstock is a major cost factor. Therefore, the use of low cost and underutilised feedstocks, such as harvest forest residues, to produce ethanol could be an interesting route. Yang and colleagues [1] evaluated several batch fermentation approaches under various conditions for ethanol production from softwood forest residues. Ranges of liquefaction time, cellulase, and yeast loadings were all evaluated in this study to improve ethanol production. A pretreatment of the lignocellulose feedstock is necessary to improve the saccharification and fermentation processes, but the current physical and/or chemical pretreatment procedures have several drawbacks. The use of laccases has been developed as an environmentally friendly alternative for improving the saccharification and fermentation stages of lignocellulosic biomass. Moreno et al. [2] evaluated a novel bacterial laccase for enhancing the hydrolysability and fermentability of steam-exploded wheat straw. To increase the productivity of the ethanol fermentation, new bioreactor designs and operation modes have been introduced. A novel textile bioreactor for improved ethanol production was developed by Osadolor and colleagues [3]. Due to the efficient mixing, this fluidised-bed bioreactor allowed the procurement of a high cell density of flocculating yeast cells, resulting in a high ethanol productivity.

Wine and beer yeasts. The increasing economic interest in the sector of sparkling wine has stimulated a renewed interest in microbial resource management. Starter cultures for sparkling wine production need to be selected in order to produce either quality base wine or to vigorously promote secondary fermentation. Garofalo and coworkers [4] reviewed the main characterisation for selecting *Saccharomyces cerevisiae* strains suitable as starter cultures and analysed the possible uses of selected non-*Saccharomyces* and malolactic strains in order to differentiate specific productions, and highlighted the main safety aspects related to microbes of enological interest. In spontaneous wine fermentations, more than one *Saccharomyces cerevisiae* strain ferments the wine must. These strains affect flavour and aroma properties differently. Therefore, Gustafsson et al. [5] investigated the interaction of two *S. cerevisiae* strains. The results showed that the co-inoculation of strains creates a new chemical profile not seen in the pure cultures, which have implications for winemakers that are looking to control wine aroma and flavour profiles through strain selection. Wine quality can be improved by using locally-selected *S. cerevisiae* strains as starter cultures. Cordero-Bueso and coworkers [6] could improve the fruity and fresh character of Malvar wines by selecting two local strains. Non-*Saccharomyces* ("wild") yeasts are found on the grapes and are also present on cellar equipment. This wealth of yeast biodiversity with hidden potential, especially for oenology, is largely untapped. In this Special Issue, the applications of non-*Saccharomyces* yeasts to the wine-making process were reviewed by Mateo and Maicas [7]. Schlander and coworkers [8] characterised acid proteases isolated from the wild yeasts *Metschnikovia pulcherrima* and *Wickerhamomyces anomalus*. These enzymes are of significant importance for medicine and biotechnology.

Yeast strains that flocculate are of particular interest to brewers, since it simplifies the yeast removal at the end of the primary fermentation considerably. Conjaerts and Willaert [9] performed adaptive laboratory evolution with gravity imposed as selective pressure for evolving a weak flocculating industrial strain towards a more flocculent phenotype. They used 3D printing to construct a suitable mini tower fermenter.

Yeast nanobiotechnology, which is a recent field where nanotechniques are used to manipulate and analyse yeast cells and cell constituents at the nanoscale, was reviewed by Willaert and coworkers [10]. An overview and discussion of nanobiotechnological analysis and manipulation techniques that have been particularly applied to yeast cells, is given; i.e., nanoscale imaging techniques, single-molecule and single cell force spectroscopy, AFM (Atomic Force Microscopy)-cantilever-based nanomotion analysis of living cells, nano/microtechniques to pattern and manipulate yeast cells, and direct contact and non-contact cell manipulation methods were reviewed.

Acknowledgments: The Belgian Federal Science Policy Office (Belspo) and the European Space Agency (ESA) PRODEX program supported this work. The Research Council of the Vrije Universiteit Brussel (Belgium) and the University of Ghent (Belgium) are acknowledged to support the Alliance Research Group VUB-UGhent NanoMicrobiology (NAMI), and the International Joint Research Group (IJRG) VUB-EPFL BioNanotechnology & NanoMedicine (NANO).

Conflicts of Interest: The author declares no conflict of interest.

References

1. Yang, M.; Ji, H.; Zhu, J. Batch Fermentation Options for High Titer Bioethanol Production from a SPORL Pretreated Douglas-Fir Forest Residue without Detoxification. *Fermentation* **2016**. [CrossRef]
2. Moreno, A.; Ibarra, D.; Mialon, A.; Ballesteros, M. A Bacterial Laccase for Enhancing Saccharification and Ethanol Fermentation of Steam-Pretreated Biomass. *Fermentation* **2016**. [CrossRef]
3. Osadolor, O.; Lennartsson, P.; Taherzadeh, M. Development of Novel Textile Bioreactor for Anaerobic Utilization of Flocculating Yeast for Ethanol Production. *Fermentation* **2015**. [CrossRef]
4. Garofalo, C.; Arena, M.; Laddomada, B.; Cappello, M.; Bleve, G.; Grieco, F.; Beneduce, L.; Berbegal, C.; Spano, G.; Capozzi, V. Starter Cultures for Sparkling Wine. *Fermentation* **2016**. [CrossRef]
5. Gustafsson, F.; Jiranek, V.; Neuner, M.; Scholl, C.; Morgan, S.; Durall, D. The Interaction of Two *Saccharomyces cerevisiae* Strains Affects Fermentation-Derived Compounds in Wine. *Fermentation* **2016**. [CrossRef]
6. Cordero-Bueso, G.; Esteve-Zarzoso, B.; Gil-Díaz, M.; García, M.; Cabellos, J.; Arroyo, T. Improvement of Malvar Wine Quality by Use of Locally-Selected *Saccharomyces cerevisiae* Strains. *Fermentation* **2016**. [CrossRef]
7. Mateo, J.; Maicas, S. Application of Non-Saccharomyces Yeasts to Wine-Making Process. *Fermentation* **2016**. [CrossRef]

8.	Schlander, M.; Distler, U.; Tenzer, S.; Thines, E.; Claus, H. Purification and Properties of Yeast Proteases Secreted by *Wickerhamomyces anomalus* 227 and *Metschnikovia pulcherrima* 446 during Growth in a White Grape Juice. *Fermentation* **2017**. [CrossRef]
9.	Conjaerts, A.; Willaert, R. Gravity-Driven Adaptive Evolution of an Industrial Brewer's Yeast Strain towards a Snowflake Phenotype in a 3D-Printed Mini Tower Fermentor. *Fermentation* **2017**. [CrossRef]
10.	Willaert, R.; Kasas, S.; Devreese, B.; Dietler, G. Yeast Nanobiotechnology. *Fermentation* **2016**. [CrossRef]

fermentation

MDPI

Article

Development of Novel Textile Bioreactor for Anaerobic Utilization of Flocculating Yeast for Ethanol Production

Osagie A. Osadolor *, Patrik R. Lennartsson and Mohammad J. Taherzadeh

Swedish Centre for Resource Recovery, University of Borås, SE 50190 Borås, Sweden;
Patrik.Lennartsson@hb.se (P.R.L.); Mohammad.Taherzadeh@hb.se (M.J.T.)
* Correspondence: Alex.Osagie@hb.se; Tel.: +46-33-435-4620; Fax: +46-33-435-4008

Academic Editor: Ronnie G. Willaert
Received: 17 September 2015; Accepted: 19 November 2015; Published: 23 November 2015

Abstract: Process development, cheaper bioreactor cost, and faster fermentation rate can aid in reducing the cost of fermentation. In this article, these ideas were combined in developing a previously introduced textile bioreactor for ethanol production. The bioreactor was developed to utilize flocculating yeast for ethanol production under anaerobic conditions. A mixing system, which works without aerators, spargers, or impellers, but utilizes the liquid content in the bioreactor for suspending the flocculating yeast to form a fluidized bed, was developed and examined. It could be used with dilution rates greater than $1.0 \, h^{-1}$ with less possibility of washout. The flow conditions required to begin and maintain a fluidized bed were determined. Fermentation experiments with flow rate and utilization of the mixing system as process variables were carried out. The results showed enhanced mass transfer as evidenced by faster fermentation rates on experiments with complete sucrose utilization after 36 h, even at 30 times lesser flow rate.

Keywords: flocculating yeast; textile bioreactor; ethanol; mass transfer; mixing; fluidization

1. Introduction

Increasing energy demand and environmental awareness have influenced the progressive rise in the production and utilization of bioethanol as a transportation fuel [1]. To boost the competitiveness of bioethanol to fossil fuel, particularly with the current low prices of fossil fuels, there is the need to continue increasing the productivity of the ethanol production process while reducing the production cost. Flocculation has proven to be advantageous in improving the productivity of the bioethanol production process [2]. Some of its benefits include production at high dilution rate, improved inhibitor tolerance, longer reuse of cells, reduced contamination tendencies at high dilution rate, reduced bioreactor cost because of smaller reactor volume [3], and ease of separating cell flocs from liquid medium in the bioreactor [4].

To effectively utilize flocculation, the size of the flocs and their settling characteristics need to be well understood. It is important that the floc size is large enough to prevent washout while being small enough to allow effective passage and mass transfer of the substrate into the cells and product out of the cells. To avoid washout, mechanical stirrers that break down the flocs are not usually used in bioreactors utilizing flocculating organisms [3]. Depending on the settling rate of the flocs, the contacting pattern that they would make in a bioreactor could result in fixed or fluidized bed. Fixed or fluidized bed systems have their benefits, but for optimal mass transfer and faster production rate, fluidized bed systems are more advantageous because of the larger contacting area of the flocs [5]. To create a stable fluidized bed, the flow rate has to be between the minimum to initiate fluidization and the maximum to prevent the flocs from being carried away from the bioreactor [5]. For these reasons, the design and operation of the bioreactor to be used for propagating flocculating microorganisms is quite important. Fluidization in bioreactors can be achieved either by aeration or by utilizing high flow liquid streams [6]. Currently, airlift bioreactors are the main type of bioreactor being used for utilizing flocculating yeast for bioethanol production [3].

Aeration is required when flocculating yeast is used in an airlift bioreactor for ethanol production [3]. This reduces the ethanol yield as ethanol is optimally produced anaerobically. Besides this, bioreactor cost is high, including aeration, generating more operation cost. Bioreactors for ethanol production have to be designed in a way that they do not hinder the activity of the microorganism within them, withstand the corrosive nature of fermentation media, and provide suitable environment and control needed to optimally produce the desired product(s) [7]. The overall goal in their design is to deliver the required functions and to be economical [8]. Conventional bioreactors for ethanol production are made using stainless steel as the major material of construction and constitute 32% of the fermentation investment cost in a typical 100,000 m^3/year plant [9]. Reducing the cost of ethanol bioreactors will reduce the cost of ethanol production. Some polymeric materials (*e.g.*, polyaniline) have good corrosion-resistance properties [10] and are cheaper than stainless steel, so they could be options for making ethanol bioreactors. However, some challenges regarding their use include their tensile strength and the effectiveness of mixing in the bioreactors made with them. A novel bioreactor with textile as its material of construction was recently introduced as a cheaper alternative to bioreactors made with stainless steel [11]. The textile bioreactor has textile as its backbone material of construction which improves its strength [11]. In this paper, the conditions needed to maintain optimal contact of the flocculating yeast in the bioreactor were determined, and the textile bioreactor was developed accordingly. This enabled anaerobic production of ethanol with the flocculating yeast, while also maintaining good mixing in the bioreactor, thus creating optimal production conditions.

2. Methods

2.1. Microorganism

Naturally flocculating yeast strain *S. cerevisiae* CCUG 53310 (Culture Collection University of Gothenburg, Gothenburg, Sweden) was used for the experiments. The flocculating yeast cells were maintained on a yeast extract peptone dextrose (YPD) agar plate containing 20 g/L agar, 20 g/L D-glucose, 20 g/L peptone, and 10 g/L yeast extract at 4 °C. Before being used for fermentation, the flocculating cells were added into 800 mL YPD media containing 20 g/L D-glucose, 10 g/L yeast extract, and 20 g/L peptone in a 2 L cotton-plugged flask. Three flasks were incubated in a shaking water bath (Grant OLS 200, Grant instrument Ltd., Cambridge, UK) at 125 rpm and 30 °C for 48 h. The supernatant liquid from two of the flasks were discarded, and the sedimented flocculating yeast were rinsed with distilled water into the textile bioreactor for fermentation. The content of the third flask was used to determine the yeast concentration in the bioreactor. A starting concentration of 2 g/L dry weight of the flocs was used for all fermentation experiments.

The average particle diameter for this yeast strain was measured using the optical density and sedimentation technique [12] and it was found to be between 190 to 320 μm. This was done by transferring samples of the cell culture to a tube in which the optical density was measured using a spectrophotometer at a wavelength of 660 nm for the various samples at different times. The optical density readings from the spectrophotometer were proportional to the flocculating yeast concentration. When all the flocculating yeast had settled (settling time), the optical density read from the spectrophotometer gave constant readings, which corresponded to the concentration of the free cells. Dividing the distance from which the flocculating yeast cells fell by the settling time gave the settling velocity. The average diameter of the flocs was calculated from Stokes' law (see equation 4), using flocculating yeast density of 1140 kg/m^3, viscosity of 0.798×10^{-3} Ns/m^2 (*i.e.*, viscosity of water at 30 °C) and acceleration due to gravity of 9.81 m/s^2.

2.2. Textile Bioreactor and Its Development

The bioreactor used for this work (Ethanolic textile lab reactor ETLRII, FOV Fabrics AB, Borås, Sweden), was a 30 L laboratory scale reactor with a working volume of 25 L. Its dimensions were 110 cm length, 8 cm depth, and 34 cm width (Figure 1a). The inlet and outlet diameter were 5 cm. It had a 4 cm opening at the middle that served as a sample collection point, a gas outlet, and a thermometer stand. Tubing was connected to the inlet and outlet of the bioreactor for recirculation.

A 12 m silicone peroxide tubing with 5 mm internal diameter and 8 mm external diameter (VWR International, Leuven, Belgium) was used as the internal mixing tubing in the bioreactor. Holes of 0.42 mm at 1 cm intervals were made in the tubing. The tubing was wound around the perimeter of the bioreactor twice and then joined into five elliptical ribbons by plastic fasteners (Figure 1b). The tubing was kept in the liquid phase by means of ten stainless steel bolts, each having internal and external diameters of 2.4 and 3.5 cm respectively, and weighing 94.6 g. The two ends of the tubing used for mixing were connected to the inlet tubing from the pump. A peristaltic pump (405U/L2 Watson-Marlow, Stockholm, Sweden) was used for recirculation, feeding, and discharging the content of the bioreactor.

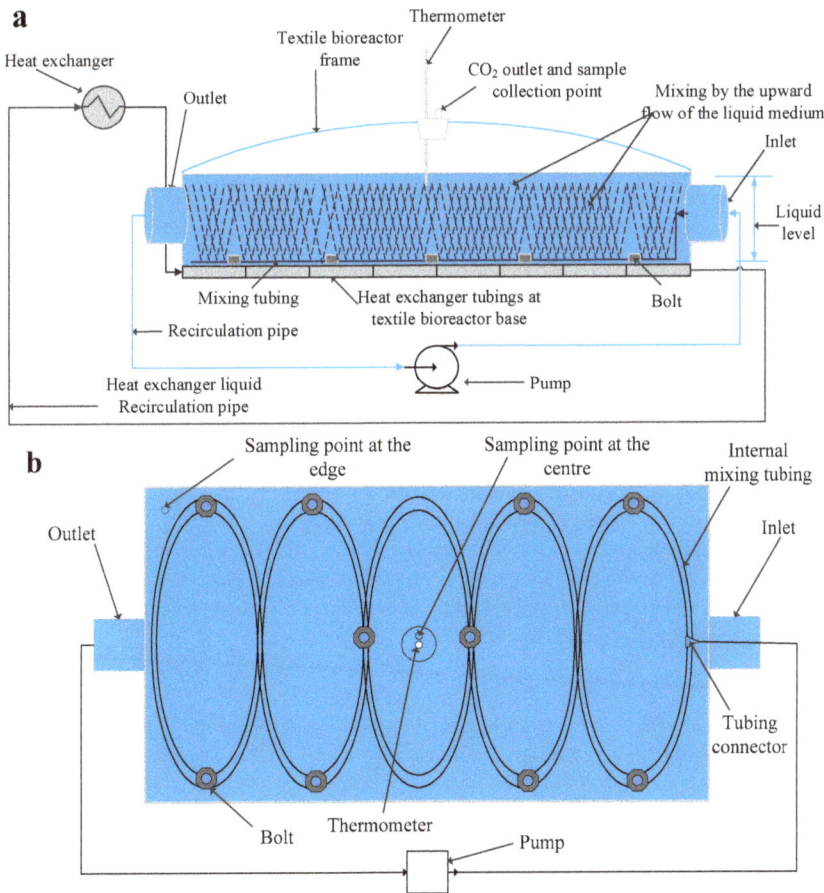

Figure 1. Developed textile bioreactor showing internal mixing inside the bioreactor from (**a**) a side view, and (**b**) a top view of the internal mixing system inside the bioreactor.

2.3. Mixing in the Bioreactor

First 20 L distilled water was fed to the bioreactor. Then, 90 ± 5 mL of 100 ppm bromophenol blue solution was added to the bioreactor at one of its ends (hereafter referred to as the injection point). The content of the bioreactor was recirculated through the internal mixing tube at flow rates of 0.002 and 0.015 volume per volume per minute (VVM). Samples were taken from the two opposite rear edges of the bioreactor. The absorbance of the samples was measured by a spectrophotometer (Biochrom Libra S60, Cambridge, UK) at 592 nm [13]. Mixing during fermentation was carried out by recirculation. To determine the effectiveness of mixing in experiments performed with internal mixing tubing, samples were taken from two sampling points, one at the center of the bioreactor and another at its edge. The samples were taken at a depth of 8 cm and from the surface of the liquid medium in the bioreactor.

2.4. Analytical Methods

Biomass was rinsed with distilled water and dried in an oven at 70 °C for 24 h, and its concentration was reported in g/L. Samples from the bioreactor were analyzed using a hydrogen-based ion exchange column (Aminex HPX-87H, Bio-road, Hercules, CA, USA) in a high performance liquid chromatography (HPLC) at 60 °C and 5 mM 0.6 mL/min H_2SO_4 eluent. A refractive index detector (Waters 2414, Waters Corporation, Milford, MA, USA) was used with the HPLC. Before being used for HPLC analysis, the samples were centrifuged for 5 min at $10,000 \times g$ and the liquid portion stored at -20 °C. Concentrations reported for all fermentation experiments were the amount determined by HPLC.

Scanning for the highest peak was used to determine the best wavelength for measuring the absorbance of bromophenol blue samples from the bioreactor. The settling rate of the flocculating yeast used for the experiment was determined using a sedimentation column. Biomass concentration of 2 g/L dry weight was released into a sedimentation column that was filled to 10 cm with the media for fermentation, and the content was mixed. The time taken for the flocs to settle to the base of the column was recorded as the settling time. The distance that the flocs settled from was divided by the settling time to give the settling velocity (V_s).

2.5. Experimental Setup for Fermentation

Fermentation experiments with and without internal mixing tubing and different flow rates (0.0016–0.06 VVM) as process variables were carried out anaerobically in the textile bioreactor. Sucrose (50–55 g/L) was used as the energy and carbon source for the flocculating yeast. It was supplemented with 1.0 g/L yeast extract, 7.5 g/L $(NH_4)_2SO_4$, 3.5 g/L KH_2PO_4, and 0.75 g/L $MgSO_4$ $7H_2O$. The liquid volume in the reactor for all experiments was 25 L. Sucrose concentration dropped between 51–45 g/L after the feed to the textile bioreactor was autoclaved. Each experiment was performed in duplicate. The error bars shown on all figures represent standard deviation values generated from the duplicated experiments.

2.6. Fluidization of the Flocs in the Bioreactor

The four forces exerted on a particle during free settling are shown in Equation (1) [14]. On the verge of fluidization, the drag force becomes equal to the pressure force acting on the flocs, and the force due to acceleration becomes zero [14]. The resulting force balance equation is shown in Equation (2). The minimum flow rate to establish this condition occurs when the superficial velocity (V_0) is equal to the fluid upwards velocity (V_u). To prevent the flocs from be carried away from the bioreactor, the maximum fluid upwards velocity (V_{umax}) should be equal to the flocs' settling velocity (V_s) [15]. For Reynolds numbers less than 10, these velocities can be obtained from Equations (3) and (4) [15].

Where ρ_p is the density of the particle, ρ_f is the density of the fluid, g is acceleration due to gravity, D_p is the particle diameter, μ is the viscosity, and ε is the void fraction.

$$\text{Force due to acceleration} = \text{Force due to gravity} - \text{Buoyancy force} - \text{Drag force} \quad (1)$$

$$\text{Pressure force } (F_p) = \text{Force due to gravity } (F_g) - \text{Buoyancy force } (F_B) \quad (2)$$

$$V_0 = V_u = (\rho_p - \rho_f)gD_p{}^2\varepsilon^3/150\,\mu\,(1 - \varepsilon) \quad (3)$$

$$V_{umax} = V_s = (\rho_p - \rho_f)gD_p{}^2/18\,\mu \quad (4)$$

For the textile bioreactor prototype used in this work, the superficial velocity is defined in Equation (5), where Q is the liquid flow rate in m^3/s and A is the surface area. The velocity of the fluid (V_i) going through the mixing tubing is defined in Equation (6), where A_i is the internal area of the tubing. Using the continuity principle, the velocity at which the fluid leaves the small holes in the tubing is 142 (*i.e.*, $2.5^2/0.21^2$) times higher than that at which it enters [15]. However, as the tubing is 12 m long and the holes are spaced 1 cm apart, there were 1200 holes. So, the upward velocity (V_h) with which the fluid emerges would split accordingly. This is shown in Equation (7). For fluidization to begin and be sustained in the developed textile bioreactor, Equation (8) shows the governing criteria:

$$V_0 = Q/A = Q/(1.1 \times 0.34) = 2.674Q \text{ (m/s)} \quad (5)$$

$$V_i = Q/A_i = Q/(\pi \times r_i{}^2) = 5.094 \times 10^4 Q \text{ (m/s)} \quad (6)$$

$$V_h = 142 \times V_i/1200 = 6.028 \times 10^3 Q \text{ (m/s)} \quad (7)$$

$$V_0 \leq V_h \leq V_s \quad (8)$$

2.7. Statistical Analysis

All statistical analyses were performed with the MINITAB® software package. Results were analyzed with ANOVA (analysis of variance), using general linear model, and factors were considered significant when they had p-value less than 0.05. The analysis was performed on the results obtained from samples measured from the start of the experiment up until the 32nd hour, when stationary phase was reached. Ethanol and sucrose concentrations were used as the response variables, the position from which the sample was taken was used as the main factor, while time and number of runs served as blocking factors.

3. Results and Discussion

Possible ways to reduce the fermentation cost in a bioethanol plant includes process development, utilizing new and cheaper bioreactors, and making the separation process more efficient. This work combines these ideas in developing the textile bioreactor. The newly developed textile bioreactor has the high flexibility, ease of operation and installation, good mechanical strength, high thermal tolerance, low purchase cost, resistance to corrosive fermentation media, and ease of transportation of its previous prototype [11]. It has a new mixing system that eliminates the need for mixing by using either axial flow impellers or aeration spargers. This removes the purchasing and operational cost associated with the maintenance of those devices. A highly flocculating yeast strain with a settling rate of 1 cm/s was used to examine the performance of the developed textile bioreactor for bioethanol production. The efficiency of mixing (a measure of the mass transfer efficiency) in the developed textile bioreactor and the flow rate needed to maintain optimal contact between the flocculating yeast and bioreactor content were investigated in this work.

3.1. Maintaining Optimal Flocs Contact in the Textile Bioreactor

Increasing the surface contact area of enzymes or catalyst in the form of a fluidized bed generally increases the speed at which a chemical reaction takes place [14]. Thus, flocculating yeast retained in a bioreactor in the form of fluidized particles would result in more rapid utilization of the substrate in the bioreactor. Using Equations (5) to (7), different fluid upflow velocities and superficial velocities for different flow rates were generated, and their values are shown in Table 1. As seen in this table, the required velocity needed to begin fluidization is low. This is because the textile bioreactor prototype used for this analysis has a high surface area to volume ratio of 12.5 m^2/m^3. The superficial velocity reduces with increasing surface area to volume ratio, so the upflow velocity needed on the verge of fluidization would be less for bioreactors having a higher surface area to volume ratio.

Table 1. The fluid upflow velocity generated at different flow rate in the textile bioreactor. VVM: volume per volume per minute; V_0: Superficial velocity; V_i: Fluid velocity; V_h: Fluid upflow velocity; V_s: Flocs settling velocity.

Q (VVM)	V_0 (m/s) $\times 10^6$	V_i (m/s)	V_h at Different Hole Spacing (m/s)			V_s (m/s)
			1 cm Spacing	5 mm Spacing	2 mm Spacing	
0.0016	1.78	0.03	0.004	0.002	0.001	0.01
0.0120	13.37	0.25	0.030	0.015	0.006	0.01
0.0160	17.83	0.34	0.040	0.020	0.008	0.01
0.0320	35.64	0.68	0.080	0.040	0.016	0.01
0.0600	66.84	1.27	0.151	0.075	0.030	0.01

For continuous production of bioethanol using flocculating yeast, it is desirable to carry out the production at a high dilution rate and at the same time prevent washout. From Equation (8), washout could occur when the fluid upflow velocity (V_h) exceeds the flocs settling velocity (V_s). From Table 1, it can be observed that washout would occur in this textile bioreactor prototype if operated on a dilution rate greater than 0.72 h^{-1} (*i.e.*, $Q \geq 0.012$ VVM), as V_h is greater than V_s. One way to increase the dilution rate would be to increase the number of holes by reducing the space between consecutive holes in the mixing tubing. From Table 1, reducing the space between the tubes to 2 mm would allow the bioreactor to be operated at a dilution rate of 1 h^{-1} with lesser tendency for washout, as V_h equals V_s with this condition. Another possibility of increasing the dilution rate would be to use mixing tubing of a length longer than 12 m. For example, using 18 m long tubing at 2 mm spacing, from equation 7, at $Q = 0.02$ VVM (dilution rate of 1.2), V_h becomes 0.007 m/s, which is less than the settling velocity of 0.01 m/s. Normally, at a dilution rate greater than the growth rate, washout would occur, but with this type of configuration for the mixing tubing, even with dilution rate higher than growth rate, washout would not occur.

3.2. Mixing as a Means of Reducing Mass Transfer Limitations

Efficient mass transfer is important for optimal performance in any bioreactor, as it helps to facilitate transfer of substrate into and product out of the microorganism, and prevent improper cell growth in the bioreactor [7]. Mass transfer is influenced by two factors: the diffusional flux between the cells and the liquid media, and the bulk flow of the liquid media. The first factor is influenced by the diffusivity of the product or substrate into the liquid media. The second is enhanced by the mixing system in the bioreactor, like agitation, aeration, or the use of stirrers [16]. With the theoretical understanding of how the flow rate can influence good contacting pattern in the previous section, experimental verifying the theoretical concepts are presented in this section.

For bioreactors utilizing flocculating yeast, mass transfer limitations need to be adequately considered. There needs to be a balance between having flocs of large sizes, favorable for the enhancement of cell retention in the bioreactor, and reducing the size of the flocs, which is necessary for the reduction of the mass transfer limitations. Characterizing solute transport into and out of the flocs is challenging because of the fragile nature of the flocs and the difficulty in deciding the geometry the flocs actually have inside the bioreactors [3]. The approach in this article was to relate the mass transfer due to diffusion to the coefficient of diffusivity, and the flow rate of the liquid media being recirculated as a measure of that due to the bulk flow of the liquid [17].

Figure 2 shows the result of the absorbance of bromophenol blue measured across the textile bioreactor at different flow rates. At 0.015 VVM, equilibrium was reached in the textile bioreactor in 15 min. At a flow rate of 0.002 VVM, equilibrium was reached in 50 min. At 25 °C, the diffusion coefficient of bromophenol blue in water is 3×10^{-10} m^2/s [18], that of ethanol in water is 1.24×10^{-9} m^2/s, that of sucrose in water is 5.24×10^{-10} m^2/s, and that of water in water is 2.45×10^{-9} m^2/s [19]. The lower diffusivity of bromophenol blue than that of the product and substrate serves as a benchmark for understanding the influence of diffusivity on mass transfer under limiting product or substrate conditions [20]. Achieving equilibrium in 50 min at a flow rate of 0.002 VVM is sufficient, as the sampling time for experiments on bioethanol production is usually measured in hours [21].

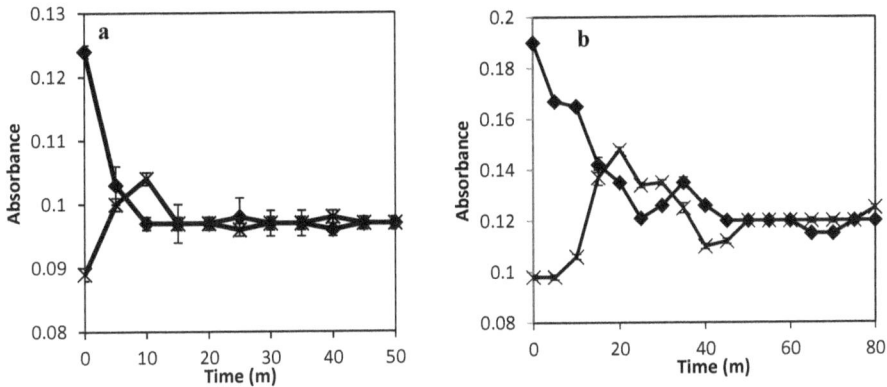

Figure 2. Bromophenol blue absorbance variation in the textile bioreactor at (**a**) a flow rate of 0.015 VVM, and (**b**) at a flow rate of 0.002 VVM at the injection point (♦) and at the opposite end (×).

3.2.1. Mass Transfer Enhancement by Internal Mixing Tubing in the Textile Bioreactor

The internal tubing had holes of 0.42 mm diameter which were 1 cm spread apart all through the textile bioreactor. Both ends of the tubing were connected to the input stream from the pump. The size of the holes and the continuous inflow of the liquid medium generate a pressure difference between the input stream and the fluid inside the tubing. A diameter of 0.42 mm was used because using a larger diameter would result in the incoming fluid leaving the tubing only from the entrance, and a smaller size would generate high back pressure on the pump.

The holes in the internal tubing create upward flow of the liquid stream, which helps in re-suspending the settling flocs, thus improving the mass transfer rate inside the textile bioreactor. To test this idea, fermentation experiments were carried out in the textile bioreactor. Figure 3a shows the result with internal tubing to aid mixing at 0.0016 VVM, while Figure 3b shows the result when mixing is not aided at a flow rate of 0.032 VVM. A lower flow rate was used for the case with aided mixing to show that the result obtained was not due to the flow rate but due to the mixing system itself—as demonstrated in Figure 2a, higher flow rate has been shown previously to improve mass transfer. The same starting concentration of 2 g/L dry weight flocculating yeast was used in both cases. For the aided mixing case (Figure 3a), the fermentation process and sucrose consumption was complete after 36 h, while the unaided mixing experiment (Figure 3b) still had more than 4 g/L of unconsumed sucrose after 48 h. The longer time in Figure 3b is attributable to the settling of the flocs at the bottom of the textile bioreactor, which causes uneven consumption of the substrate in the reactor. The superficial velocity for Figure 3b from Table 1 is 3.6×10^{-5} m/s, which can be met by the velocity of the CO_2 gas bubbles rising from the bottom of the reactor, at the onset of the fermentation experiment. This could explain the higher substrate consumption during the first 28 h of fermentation than in the latter part. However, when the sucrose at the bottom of the reactor is consumed, there will no longer be CO_2 bubbles to re-suspend the settled flocs, so the flocs at the bottom of the reactor, because of the mass transfer limitation, would go into the stationary phase faster, resulting in longer fermentation time or incomplete utilization of the substrate. In Figure 3a, the upward flow of the liquid stream helped to keep the flocs uniformly distributed in the textile bioreactor through the duration of the fermentation experiment. This shows that the developed mixing system for the textile bioreactor is effective in preventing ineffective substrate utilization and increasing the fermentation rate. From Figure 3a, the average peak ethanol concentration was 22.13 ± 0.93 g/L. Using the average fermentation time gave the specific ethanol productivity in the developed textile bioreactor with aided mixing at 0.0016 VVM as 0.29 ± 0.01 g-ethanol/g-biomass/h. The best specific productivity from a gas lift reactor with recycle was reported as 0.045 g-ethanol/g-biomass/h with no sugar loss, and 0.43 g-ethanol/g-biomass/h with significant sugar loss [22]. For an airlift reactor, optimum specific productivity of 0.4 g-ethanol/g-biomass/h with significant sugar loss has been reported [23]. Comparing the reported specific productivity of different bioreactors for ethanol production from literature with that of the developed textile bioreactor, it can be seen that for an optimal combination of high specific productivity and complete sugar utilization, the developed textile bioreactor performs better.

3.2.2. Higher Flow Rate with and without Internal Tubing

The mass transfer rate can be enhanced to an extent by increasing the bulk flow of the liquid media [16]. The effect of higher flow rate on the fermentation rate was investigated. For the textile bioreactor with internal mixing tubing, a flow rate of 0.012 VVM (7.5 times higher than the previously examined flow rate) was used as the recirculation rate in the textile bioreactor with internal mixing tubing. Figure 4a shows the result of this experiment. From Figure 4a, the fermentation rate was slightly faster in the first 32 h of fermentation in comparison to the experiment with a recirculation rate of 0.0016 VVM. However, the substrate was fully consumed after 36 h in both cases. This shows that mixing is not the rate limiting step in both cases, but rather sucrose hydrolysis and slow fructose utilization [22]. This limitation can be handled in several ways such as operating the reactor in a fed-batch mode [24], or by maintaining a high concentration of flocs in the bioreactor [22]. However, for continuous production, using a flow rate of 0.012 VVM at 1 cm hole spacing would cause washout to occur much faster than that at a flow rate of 0.0016 VVM (Table 1). This did not affect the fermentation rate in the experiment performed at a flow rate of 0.012 VVM because the cells were recycled back into the textile bioreactor.

For the case without the mixing tubing, a flow rate of 0.06 VVM (1.9 times higher than the previously examined flow rate) was used as the recirculation rate. The result of this experiment is shown in Figure 4b. Comparing the experiment performed at 0.06 VVM with that at 0.032 VVM shows that the higher recirculation flow rate increased the fermentation rate. However, the higher flow rate was not sufficient to cause complete utilization of the sucrose within 48 h because the settling rate of the flocs is higher than the turbulence created by the high flow rate. Comparing the experiment performed at 0.0016 VVM with that performed at 0.06 VVM shows that the developed mixing system is much more effective in overcoming mass transfer limitations associated with mixing, even with 30 times slower flow rate.

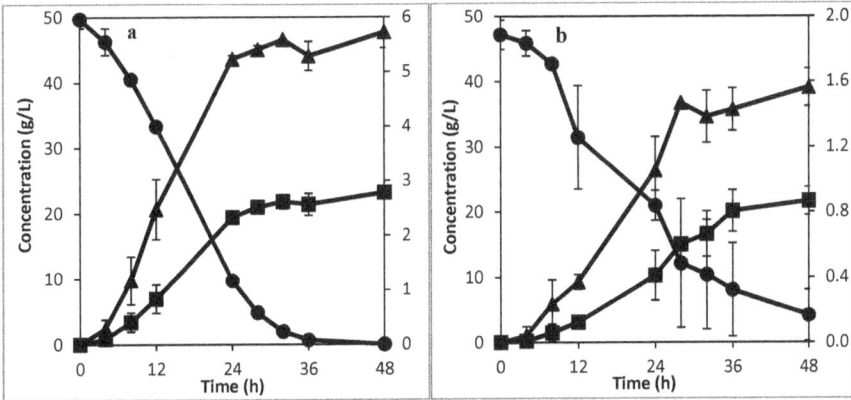

Figure 3. Experiment with enhanced mixing at (**a**) a recirculation rate of 0.0016 VVM , and (**b**) without enhanced mixing at 0.032 VVM, showing ethanol (■) and sucrose (●) concentration in the primary axis (right hand side), and glycerol concentration (▲) in the secondary axis.

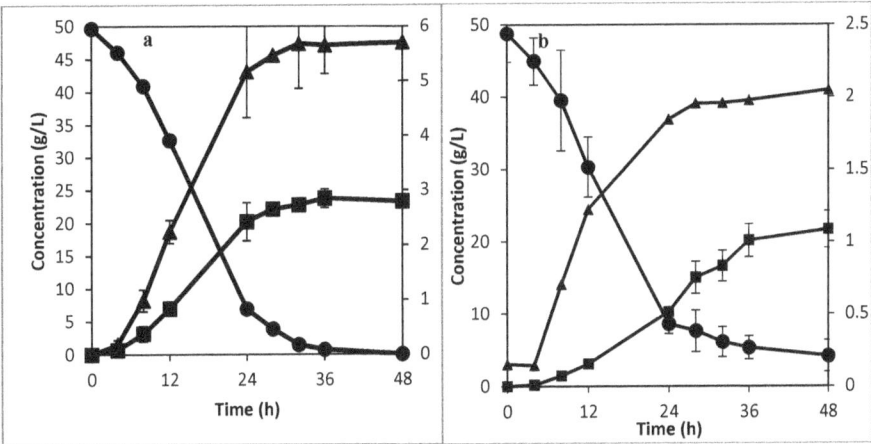

Figure 4. Experiments with (**a**) enhanced mixing at 0.012 VVM, and (**b**) without enhanced mixing at a recirculation rate of 0.06 VVM , showing ethanol (■) and sucrose (●) concentration in the primary axis (right hand side), and glycerol concentration (▲) in the secondary axis (left hand side).

3.3. Mixing along the Edges of the Textile Bioreactor

One of the limitations of a rectangular reactor is the possibility of it having poor mixing, especially at the edges [25]. Ethanol and sucrose concentrations across different sampling positions in the developed textile bioreactor at a recirculation rate of 0.0016 VVM are shown in Table 2. From Table 2 it can be seen that the concentrations of ethanol and sucrose from different sampling points and depths are similar. ANOVA using a general linear model of ethanol and sucrose concentration as responses and sampling position as factor gave p-values of 0.861 for ethanol and 0.733 for sucrose, so the position from which the sample is drawn is not statistically significant, meaning that the mixing is uniform at all points in the textile bioreactor. This uniformity in the textile bioreactor implies that the flocs responsible for the conversion of substrate to product is uniformly distributed across the developed textile bioreactor. From this it can be seen that the developed textile bioreactor does not have regions with significant ethanol or sucrose concentration variation, so there will be efficient and fast utilization of the substrate across it.

Table 2. Ethanol and sucrose concentrations from different sampling point in the textile bioreactor.

Time (h)	Run	Ethanol Concentration (g/L)				Sucrose Concentration (g/L)			
		Edge 8 cm Deep	Edge Surface	Centre 8 cm Deep	Centre Surface	Edge 8 cm Deep	Edge Surface	Centre 8 cm Deep	Centre Surface
0	1	0.00	0.00	0.00	0.00	48.96	48.46	48.75	48.75
4	1	0.86	0.45	0.51	0.47	46.22	44.97	44.84	44.84
8	1	2.39	2.30	2.15	2.38	43.61	41.91	40.97	40.97
12	1	5.14	5.41	5.23	5.44	36.16	38.04	36.01	33.70
24	1	17.91	18.24	18.87	18.85	11.61	5.52	12.11	9.34
28	1	20.46	19.95	20.60	20.44	3.21	2.13	3.14	5.50
32	1	21.20	21.25	21.68	21.04	1.35	1.35	1.38	1.39
0	2	0.00	0.00	0.00	0.00	48.19	50.40	50.68	50.68
4	2	1.18	1.18	1.27	1.21	46.86	46.86	47.72	47.72
8	2	4.56	4.41	4.41	4.44	40.87	39.94	40.10	40.10
12	2	8.60	8.27	8.23	8.53	32.41	30.77	31.84	32.98
24	2	20.03	19.94	20.01	19.98	10.00	10.14	10.03	10.16
28	2	21.93	21.65	21.88	21.52	4.34	4.16	4.31	4.22
32	2	22.60	22.05	22.65	22.59	2.30	2.82	2.87	2.55

3.4. Ethanol Production Process Development and Cost Reduction

Ethanol could be produced by batch, fed-batch, or continuous modes of production. Conventionally, ethanol is produced anaerobically, and the yeast cells are centrifuged and either recycled or channeled somewhere else [26]. However, for a 100,000 m^3/year ethanol production facility, the centrifuge accounts for 18% of the fermentation investment cost [9], excluding the energy cost. This can be eliminated by using flocculating yeast. Mixing in bioreactors using flocculating yeast is carried out using aeration (so the bioreactor is operated as a bubble column or an airlift reactor) as stirrers breaks the flocs [3]. Despite the Crabtree effect that favors ethanol production by *S. cerevisiae* under aerobic conditions [27], aerobic conditions cause more biomass production, which reduces the amount of ethanol that can be produced [28]. The developed textile bioreactor has been shown to both anaerobically use flocculating yeast for ethanol production and maintain good mixing, thus it can combine the benefits of optimal ethanol production and elimination of centrifugation cost.

Using the same volume of textile bioreactor in place of a stainless steel reactor has been previously shown to reduce the fermentation investment cost by 19% [11]. Combining this with the savings obtained by not using a centrifuge would result in 37% reduction in the investment cost of a 100,000 m^3/year ethanol production facility [9].

4. Conclusions

The developed textile bioreactor showed a better combination of specific ethanol productivity and complete sugar utilization than both gas lift bioreactors with recycle and airlift bioreactors when using flocculating yeast for ethanol production. The results of the anaerobic mixing efficiency with high settling flocculating yeast showed excellent mixing in comparison with the previous prototype. This makes it possible to combine optimal anaerobic ethanol production rates and the process benefits of using flocculating yeast for ethanol production. Additionally, for yeast flocs having average particle diameter between 190 to 320 μm, there is the possibility of operating the bioreactor at a dilution rate of more than 1 h^{-1} with less chances of washout by using mixing tubing configurations that makes the fluid upflow velocity less than the flocs settling velocity. Using flocculating yeast as the fermenter in the same volume of the developed textile bioreactor as the conventionally used bioreactor in a 100,000 m^3/year ethanol production facility can give a 37% reduction in fermentation investment cost.

Fermentation **2015**, *1*, 98–112

Acknowledgments: The authors would like to appreciate Magnus Lundin, Peter Therning, and Regina Patinvoh for their technical assistance and valuable discussions. The textile bioreactor used for this article was kindly provided by FOV Fabrics AB (Borås, Sweden).

Author Contributions: Osagie A. Osadolor did the experiment design, performed all the experiment, wrote most of the manuscript and had the idea. Patrik R. Lennartsson was responsible for part of the manuscript. Mohammad J. Taherzadeh was responsible for part of the manuscript. All authors have given their approval for this final version of the manuscript.

Conflicts of Interest: The authors declare no conflict of interest.

References

1. Balat, M.; Balat, H. Recent trends in global production and utilization of bio-ethanol fuel. *Appl. Energ.* **2009**, *86*, 2273–2282. [CrossRef]
2. Mussatto, S.I.; Dragone, G.; Guimarães, P.M.R.; Silva, J.P.A.; Carneiro, L.M.; Roberto, I.C.; Vicente, A.; Domingues, L.; Teixeira, J.A. Technological trends, global market, and challenges of bio-ethanol production. *Biotechnol. Adv.* **2010**, *28*, 817–830. [CrossRef] [PubMed]
3. Domingues, L.; Vicente, A.A.; Lima, N.; Teixeira, J.A. Applications of yeast flocculation in biotechnological processes. *Biotechnol. Bioprocess. Eng.* **2000**, *5*, 288–305. [CrossRef]
4. Verstrepen, K.J.; Klis, F.M. Flocculation, adhesion and biofilm formation in yeasts. *Mol. Microbiol.* **2006**, *60*, 5–15. [CrossRef] [PubMed]
5. Ergun, S.; Orning, A.A. Fluid flow through randomly packed columns and fluidized beds. *J. Ind. Eng. Chem.* **1949**, *41*, 1179–1184. [CrossRef]
6. Muroyama, K.; Fan, L.S. Fundamentals of gas-liquid-solid fluidization. *AIChE. J.* **1985**, *31*, 1–34. [CrossRef]
7. Blakebrough, N. Fundamentals of fermenter design. *Pure Appl. Chem.* **1973**, *36*, 305–316. [CrossRef]
8. Mark, R.; Wilkins, A.H.A. Fermentation. In *Food and Industrial Bioproducts and Bioprocessing*; John Wiley & Sons: Hoboken, NJ, USA, 2012.
9. Maiorella, B.L.; Blanch, H.W.; Wilke, C.R.; Wyman, C.E. Economic evaluation of alternative ethanol fermentation processes. *Biotechnol. Bioeng.* **2009**, *104*, 419–443. [CrossRef] [PubMed]
10. Wei, Y.; Wang, J.; Jia, X.; Yeh, J.-M.; Spellane, P. Polyaniline as corrosion protection coatings on cold rolled steel. *Polymer* **1995**, *36*, 4535–4537. [CrossRef]
11. Osadolor, O.A.; Lennartsson, P.R.; Taherzadeh, M.J. Introducing textiles as material of construction of ethanol bioreactors. *Energies* **2014**, *7*, 7555–7567. [CrossRef]
12. Van Hamersveld, E.H.; van der Lans, R.; Luyben, K. Quantification of brewers' yeast flocculation in a stirred tank: Effect of physical parameters on flocculation. *Biotechnol. Bioeng.* **1997**, *56*, 190–200. [CrossRef]
13. Nagai, Y.; Unsworth, L.D.; Koutsopoulos, S.; Zhang, S. Slow release of molecules in self-assembling peptide nanofiber scaffold. *J. Control. Release* **2006**, *115*, 18–25. [CrossRef] [PubMed]
14. Kunii, D.; Levenspiel, O. *Fluidization Engineering*; Elsevier: Amsterdam, The Netherlands, 2013.
15. McCabe, W.L.; Smith, J.C.; Harriott, P. *Unit Operations of Chemical Engineering*; McGraw-Hill: New York, NY, USA, 1993; Volume 5.
16. Benitez, J. *Principles and Modern Applications of Mass Transfer Operations*; John Wiley & Sons: Hoboken, NJ, USA, 2011.
17. Bergman, T.L.; Incropera, F.P.; Lavine, A.S. *Fundamentals of Heat and Mass Transfer*; John Wiley & Sons: Hoboken, NJ, USA, 2011.
18. West, J.; Gleeson, J.P.; Alderman, J.; Collins, J.K.; Berney, H. Structuring laminar flows using annular magnetohydrodynamic actuation. *Sensor Actuator B* **2003**, *96*, 190–199. [CrossRef]
19. Hayduk, W.; Laudie, H. Prediction of diffusion coefficients for nonelectrolytes in dilute aqueous solutions. *AIChE. J.* **1974**, *20*, 611–615. [CrossRef]
20. Bosma, T.N.P.; Middeldorp, P.J.M.; Schraa, G.; Zehnder, A.J.B. Mass transfer limitation of biotransformation: Quantifying bioavailability. *Environ. Sci. Technol.* **1996**, *31*, 248–252. [CrossRef]
21. Elander, R.T.; Putsche, V.L. *Ethanol from Corn: Technology and Economics*; Taylor and Francis: Washington, DC, USA, 1996; pp. 329–349.
22. Fontana, A.; Ghommidh, C.; Guiraud, J.P.; Navarro, J.M. Continuous alcoholic fermentation of sucrose using flocculating yeast. The limits of invertase activity. *Biotechnol. Lett.* **1992**, *14*, 505–510. [CrossRef]

23. Sousa, M.L.; Teixeira, J.A.; Mota, M. Comparative analysis of ethanolic fermentation in two continuous flocculation bioreactors and effect of flocculation additive. *Bioprocess. Eng.* **1994**, *11*, 83–90. [CrossRef]
24. Echegaray, O.F.; Carvalho, J.C.M.; Fernandes, A.N.R.; Sato, S.; Aquarone, E.; Vitolo, M. Fed-batch culture of saccharomyces cerevisiae in sugar-cane blackstrap molasses: Invertase activity of intact cells in ethanol fermentation. *Biomass Bioenergy* **2000**, *19*, 39–50. [CrossRef]
25. Oca, J.; Masaló, I. Design criteria for rotating flow cells in rectangular aquaculture tanks. *Aquacult. Eng.* **2007**, *36*, 36–44. [CrossRef]
26. Basso, L.C.; Rocha, S.N.; Basso, T.O. *Ethanol Production in Brazil: The Industrial Process and its Impact on Yeast Fermentation*; Intech: Rijeka, Croatia, 2011.
27. Verduyn, C.; Zomerdijk, T.P.L.; van Dijken, J.P.; Scheffers, W.A. Continuous measurement of ethanol production by aerobic yeast suspensions with an enzyme electrode. *Appl. Microbiol. Biotechnol.* **1984**, *19*, 181–185. [CrossRef]
28. Taherzadeh, M.J.; Lennartsson, P.R.; Teichert, O.; Nordholm, H. Bioethanol production processes. *Biofuels Prod.* **2013**, 211–253. [CrossRef]

fermentation

MDPI

Article

Improvement of Malvar Wine Quality by Use of Locally-Selected *Saccharomyces cerevisiae* Strains

Gustavo Cordero-Bueso [1,2,*], Braulio Esteve-Zarzoso [3], Mar Gil-Díaz [1], Margarita García [1], Juan Mariano Cabellos [1] and Teresa Arroyo [1]

[1] Departamento de Agroalimentación, Instituto Madrileño de Investigación y Desarrollo Rural Agrario y Alimentario, Autovía A2 km 38.2, Alcalá de Henares, Madrid 28800, Spain; mar.gil.diaz@madrid.org (M.G.-D.); margarita_garcia_garcia@madrid.org (M.G.); juanmariano.cabellos@madrid.org (J.M.C.); teresa.arroyo@madrid.org (T.A.)
[2] Departamento de Biomedicina, Biotecnología y Salud Pública, Universidad de Cádiz, Avenida de la República Saharaui s/n, Puerto Real, Cádiz 11510, Spain
[3] Biotecnologia Enològica, Departament de Bioquimica i Biotecnologia, Facultat d'Enologia, Universitat Rovira i Virgili, Marcel li Domingo s/n, Tarragona 43007, Spain; Braulio.esteve@urv.cat
* Correspondence: gustavo.cordero@uca.es; Tel.: +34-956-016-424

Academic Editor: Ronnie G. Willaert
Received: 16 February 2016; Accepted: 10 March 2016; Published: 14 March 2016

Abstract: Malvar grape juice offers relatively little in the way of a sensory experience. Our interest lies in the use of locally-selected yeast strains in experimental fermentations to improve the sensory characteristics of Malvar wines. Two locally-selected strains of *Saccharomyces cerevisiae* were used as starter cultures in vinifications and compared with spontaneous fermentations of the same cultivar musts. Wine quality was investigated by their principal oenological parameters, analysis of the volatile aroma components, and corroborated by an experienced taster panel. The most salient chemical attributes were its high concentrations of isoamyl acetate and hexyl acetate and the high acidity, which have been detected to be key constituents in setting the fruity and fresh character of Malvar wines. Winemakers of winegrowing areas where this grape variety is cultivated will have improved options to elaborate new white wines styles, using selected yeast strains that enhance its aromatic properties.

Keywords: sensory analysis; inoculation; spontaneous fermentation; yeast selection

1. Introduction

One of the most cultivated varieties in the winegrowing region of Madrid, with a total extension of 11,758 ha, is the white grape variety Malvar (*Vitis vinifera* L.). Part of the economic development of this area is based on wine production using grapes from this cultivar. However, Malvar grape juice offers relatively little in the way of a sensory experience, but after wine fermentation, it has a multitude of attributes [1]. Hence, the intervention of the yeasts in the fermentation process can be noteworthy. In a previous work, we characterized the enological relevance of 12 selected non-*Saccharomyces* wine yeast species isolated from spontaneous fermentations of this variety to propose their use in mixed starter cultures to improve the organoleptic properties of the Malvar wines [1]. The yeast strain *Torulaspora delbrueckii* CLI 918 was defined as a yeast strain with potential interest for its contribution to the aromatic Malvar wine profile with flowery and fruity aromas. This yeast species could be used in mixed starter cultures with *S. cerevisiae*. It is well known that ethanol-tolerant yeasts, such as *S. cerevisiae*, play an essential role in the evolution of yeast-derived and grape-derived flavor molecules [2–6]. However, not all wine yeasts strains are equal; juice from the same grapes will deliver quite different wines, depending on the choice of yeast [7] and the management of the fermentation processes [8]. Thus, revealing new data relevant to Malvar wines is necessary.

Autochthonous yeast strains selected for their use as starter cultures is a profitable approach. Thus, lately, in a challenge to enrich distinctive aromatic properties, some researchers have addressed their attention to the selection of yeasts from restricted areas [8–12]. Ecological studies of wine yeasts are essential for finding novel strains with new molecular and enological attributes. Some winemakers proclaim that wines with geographical characteristics can be obtained only with selected yeast starters originated from the same area where wines are elaborated and the use of autochthonous wine yeasts selected from each vine-growing area is widely spread [8,9].

Uninoculated fermentation is a complex microbial process accomplished by the sequential action of non-*Saccharomyces* and the different *Saccharomyces* yeast strain populations present on the skin of the grapes, winery equipment, in the musts, and in the wine [13]. On the other hand, inoculation with indigenous yeasts as a starter culture is engaged to set a dominant population of a selected strain from the beginning to the end of fermentation [14,15]. The yeast activity during a spontaneous fermentation is capricious and could contribute less desirable attributes to the wine. Moreover, risks related with natural fermentations include both sluggish or arrested fermentations and the propagation of contaminant yeasts [13,14]. To keep away of these problems, commercially available wine yeast exhibits great diversity in degrees of robustness to dryness but, unfortunately, the most resilient strains do not necessarily deliver optimal sensory characteristics to the wine [7,15]. In addition, the use of commercial starters could disguise the distinctive properties that characterize some local wines [7,15]. Thus, understanding how yeasts influence the principal properties of wine aroma, flavor, and color provided the basic steps for selection of autochthonous yeast strains for use as starter cultures and control of the alcoholic fermentation as a new commercial option for wine makers. Furthermore, the use of selected native yeast strains in starter cultures is rather preferred since they are better acclimatized to the environmental circumstances and may ensure the maintenance of the typical sensory characteristics of the wines of a certain region [11].

The goal of this study is the evaluation of Malvar wines made with indigenous and selected yeast strains of *S. cerevisiae* in order to evaluate the effect, to devise the use of these yeast strains to make wine, and to determine the most remarkable chemical and sensory characteristics of such wines belonging to the Appellation of Origin "Vinos de Madrid" (Madrid, Spain). Moreover, no investigations have been carried out to improve its enological characteristics using selected or autochthonous strains of *S. cerevisiae* selected in this area.

2. Experimental Section

2.1. Yeast Strains and Vinification Procedure

The yeast strains utilized in this study were two *S. cerevisiae* yeast strains (coded as CLI 889 and CLI 892) previously isolated among 18 different genetic profiles obtained by PFGE with different occurrence in the Madrid winegrowing region. These were selected and characterized in our laboratories based on some established and desirable enological criteria, such as high fermentation performance, resistance to ethanol, low production of hydrogen sulfide and sulfur dioxide, and volatile-derived, among others [16] (see Supplementary Material). Fermentations were conducted at the IMIDRA's experimental cellar. Must obtained showed 240 g/L of reducing sugars, 170 mg/L of yeast absorbable nitrogen (YAN), 900 mg/L of total amino acids, the pH value was 3.61, and the titratable acidity (expressed as g/L of tartaric acid) was 5.78. Musts were carefully racked, homogenized, and dislodged statically (at 4 °C) adding 0.01 g/L of pectolytic enzymes (Enozym Altair, Agrovin, Spain) and 50 mg/L of sulfur dioxide (SO_2). This must was divided into nine stainless steel vats of 100 L coded as A1, A2, and A3 (those inoculated with CLI 889); B1, B2, and B3 (those inoculated with CLI 892), and the spontaneous fermentations as S1, S2, and S3. Triplicate fermentations were carried out at a controlled temperature of 18 °C. Musts were inoculated with a final concentration of 10^6 cells/ml of pure selected yeast. Fermentation kinetic was controlled by monitoring daily the density. When its value was the same during two consecutive days, residual sugars were analyzed by enzymatic methodology (Roche diagnostics, Darmstadt, Germany). Fermentation was considered to be completed when residual sugars concentration was less than 2 g/L. After fermentation, the wines were clarified by cold settling using 3 mL/L of colloidal silica (Silisol, Agrovin, Alcázar de San Juan, Ciudad Real, Spain) and 2 mg/L of gelatin fining (Vinigel, Agrovin, Alcázar de San Juan, Ciudad Real, Spain). After three months of cold stabilization at 4 °C, wines were filtered (0.6 μm) and bottled. Then, they were subjected to chemical analysis.

2.2. Yeast Isolation, Identification, and Typification

Samples were taken from every vat during the vinification process at different density values, initial (D1) = 1090 g/L, D2 = 1085–1070 g/L, D3 = 1060–1050 g/L, D4 = 1030–1025 g/L, D5 = 1010–1000 g/L and D6 = 990 g/L (<2 g/L of residual sugars of the fermentations). Thus, fifty-four 100 mL sterile plastic flasks were filled with the Malvar must/wine from different parts of the vessels, kept under refrigeration (4 °C), and transported to the laboratory. Aliquots of tenfold dilution of the samples were spread onto YGC agar plates (Laboratorios Conda, Madrid, Spain). The plates were incubated for 3–4 days at 26 °C. After yeast colony counting, 30 colonies were randomly selected from each fermentation sample for their identification and subsequent monitoring of the implantation rate.

DNA extraction and quantification from isolates was performed as stated by Cordero-Bueso *et al.* [17]. Identification was carried out by the amplification of the ITS1-5.8S-ITS2 r DNA region and subsequent RFLP analysis using endonucleases *Cfo*I, *Hinf*I, *Hae*II, and *Dde*I [18]. Those isolated and identified different to *S. cerevisiae* were not analyzed in this study. In order to monitor the yeast strain dominance during the fermentation processes, two methodologies were used, karyotyping by pulsed field gel electrophoresis (PFGE) and amplification by PCR of microsatellite regions. The molecular karyotype was obtained following the protocol proposed by Rodríguez *et al.* [14]. Microsatellite reaction mix and amplification protocols were identical as those used by Vaudano and García-Moruno [19]. Amplified products were scattered on an agarose gel (2.5% *w/v*) with a final concentration of 5 μL/mL of ethidium bromide, in 1× TBE buffer at 100 V for 90 min. DNA fragment sizes were resolved by comparison with a molecular ladder marker of 100 bp (Promega, Madison, WI, USA). Fragment differentiation and allele size determination was carried out by single capillary automatic electrophoresis (CE) in ABI 3130 Genetic Analyzer (Applied Biosystems, Foster City, CA, USA).

2.3. Enological Parameters of the Fermentation Assays

Enological parameters measurements were performed as stated in Cordero-Bueso *et al.* [1]. Alcoholic titer was measured by near-infrared reflectance, the fermentative capacity was calculated as the difference between the initial and final sugar content. Fermentation velocity (V_F) (or alcohol production expressed as grams of sugar consumed daily) was measured checking daily the sugar percentage lost during the fermentation. In addition, V_{50} amount of sugar daily transformed by the yeasts when 50% of the sugar content had been used up was evaluated. Free and total sulfur dioxide, pH, titratable acidity, and volatile acidity were performed according to standard methods in the enological sector (OIV methods, Official Methods established by the European Union). Glycerol and 2,3-butanediol compounds were determined using the Feuilles verts 588 (FV) method also suggested by the OIV.

2.4. Determination of Carboxylic Acids of the Malvar Wines

Carboxylic acids were quantified by ionic chromatography using Dionex DX 500 (Salt Lake City, UT, USA) equipment with a CD20 conductivity detector. Standard stock solutions of the organic acids (Panreac, Barcelona, Spain) were prepared by dissolution of the acids or the salts with deionized water. After filtering (0.22 μm) and dilution (1:20) with sterilized water, duplicate wine samples were injected into the chromatograph equipped with an IonPac ICE-AS6 capillary column. A concentration of 0.4 mM heptafluorobutyric acid (HFBA) (FlukaChemie AG, Buchs, Switzerland) was used as eluent at flow rate of 1.0 mL/min in isocratic mode. Anion-Ice micro-membrane was used as suppressor column and tetrabutylamonium hydroxide (Riedel-de Haën, Seelze, Germany) as a regenerator with a flow rate of 5 mL/min. The working conditions were as follow; temperature at 25 °C, 25 μL of injection volume, and 10 μs FS of the detector conductivity.

2.5. Determination of the Volatile Fraction of the Fermentations

Six major volatile compounds of the Malvar wines obtained were settled by gas chromatography coupled to a flame ionization detector (GC-FID) and 19 minor volatiles by gas chromatography combined with mass spectrometry (GC-MS). Major volatiles (acetaldehyde, acetoin, ethyl acetate, and the higher major alcohols, 1-propanol, isobutanol, and the isoamylic alcohols (2-methyl-1-butanol and 3-methyl-1-butanol)) were determined after steam distillation using a Hewlett Packard Series II (Palo Alto, CA, USA) gas chromatograph with a flame ionization detector. At the same time, minor volatiles were extracted after a liquid–liquid process using a mixture of ether hexane (1:1 v/v) as extractant. The organic phases were concentrated under a stream of N_2 and injected into a Hewlett-Packard 6890 gas chromatograph (Agilent, Avondale, PA, USA) fitted to a mass spectrometer detector HP Mass Selective 5973 (Agilent, Avondale, PA, USA). The reference Gil *et al.* [20] shows a complete description of both methods.

2.6. Sensory Analysis

Sensory evaluations were done under ISO standards [21–23] related to methodology and sensory analysis vocabulary (ISO 8586-1:1993), selection and formation of tasters (ISO 11035:1994), and tasting room (ISO 8589:2007). Two sensory analyses were performed in only one session by eleven skilled judges. Malvar wine sample positions were randomized every time and the sensory profile was defined using 13 descriptors previously described by Lozano *et al.* [24] and chosen by the taster panel in a previous session as stated in the ISO 11035:1994 rules and according to their importance in Malvar wines.

The first wine tasting was carried out by filling in a blank official tasting scorecard used in the Appellation of Origin "Vinos de Madrid". Penalizing scores were used; thus, the better quality wines obtained a lower score. Six variables (appearance, aroma quality, aroma intensity, taste intensity, taste quality, and harmony) were selected for estimation of wine quality, and a scale of seven categories (excellent: 0–7, very good: 8–23, good: 24–44, correct: 45–52, ordinary: 53–78, defective: 79–90, eliminated > 90) like those proposed by Vilanova *et al.* [25]. The total scores given by the eleven tasters for each parameter corresponding to the sensorial characteristics of wine were then statistically analyzed. After the first wine tasting, all judges were also asked to evaluate the sensory profile of the samples on a 0–5 point scale of intensity filling in a second official tasting scorecard containing the aroma descriptors mentioned above. Scale zero (0) implied that the descriptor was not perceived, while a score of five (5) was equal to the highest perception.

2.7. Statistical Analysis

One-way ANOVA and Tukey's test were accomplished to emphasize the effects of yeast strains on sensorial descriptors. Discriminant analysis was performed to point up any differences due to yeast strain. Principal component analysis (PCA) was carried out with 20.0 SPSS (Inc. Chicago, IL, USA) for Windows statistical package (significance level $p = 0.05$).

3. Results and Discussion

3.1. Fermentation Kinetics

Figure 1 shows the average of the fermentative kinetics and yeast population evolution of Malvar musts. A lag phase of two days was observed in all fermentations and proceeding to dryness between eight and nine days in the case of the inoculated fermenters, and between 10 and 11 days in the spontaneous fermentations. Although different kinetic behaviors were found, all strains were able to finish the fermentation consuming over 98.5% of initial sugar. Differences on sugar consumption in the middle stages of the fermentations have been observed. In spite of the similar profile obtained for the inoculated fermentations during the first days (until the third day) and at the completion of fermentation (from the sixth day), the strain CLI 892 showed higher fermentation rate than the CLI 889 strain and spontaneous fermentations, which showed the same rate during this period of time.

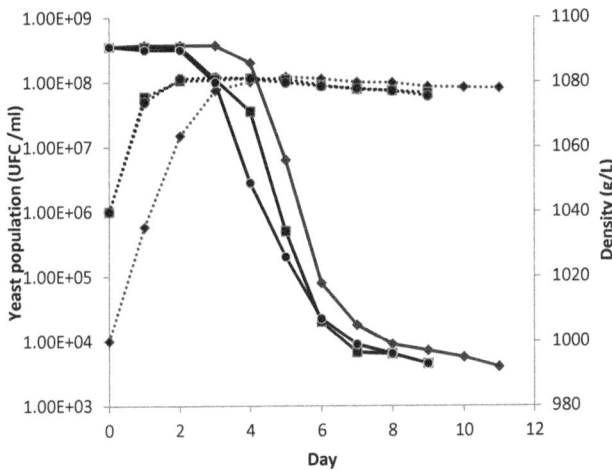

Figure 1. Evolution of the total yeast population (dashed lines) and must density (continuous lines) during three wine fermentations performed with Malvar musts using the following inoculation procedures: no inoculation (◆); inoculation with yeast strain CLI 889 (■); and inoculation with yeast strain CLI 892 (●). Data correspond to mean values obtained from triplicate experiments.

3.2. Implantation Rate

A total of 1620 colonies were isolated from the different stages of the fermentations of Malvar among the nine fermenters. Molecular identification using ITS-5.8S amplification and restriction analysis, and comparing the restriction profiles with those obtained by Esteve-Zarzoso *et al.* [18], showed that from first day of inoculated fermentations all isolates belonged to *S. cerevisiae*, while in the uninoculated musts, 131 of 540 colonies were identified as non-*Saccharomyces* in the initial and middle stages of the spontaneous fermentations.

All colonies identified as *S. cerevisiae* were characterized by PFGE, showing five different karyotypes. Isolates from inoculated vessels with the yeast strain CLI 889 showed two different karyotype patterns during the early and middle phases of the fermentations, but at the end of the fermentation process only the karyotype (A) corresponding to the inoculated yeast strain was found (Figure 2), while colonies of *S. cerevisiae* from the three vessels inoculated with the yeast strain CLI 892 showed one unique profile (B), throughout the entire fermentation process and corresponding to the inoculated yeast (Figure 2). The implantation rate at the end of the fermentation 100% fit the profile of the inoculated strain in both tanks. While in the uninoculated vessels, 90% of the population was represented by a single karyotype (A) as showed Figure 2. Dominance or competitiveness of a starter yeast strain could have an impact on the sensorial quality of wine by dominating its aromatic profile or eliminating the collaborative role of natural *S. cerevisiae* populations [13].

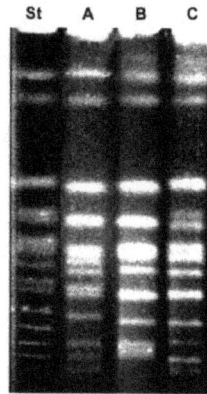

Figure 2. Electrophoretic karyotypes of the majority at the end of the fermentation *Saccharomyces cerevisiae* yeast strain isolated from the spontaneous fermentations (**A**) and selected inoculated yeast strains CLI 889 (**B**) and CLI 892 (**C**) The chromosomes of the standard *S cerevisiae* YNN 295 (Bio-Rad) were used as a reference (St).

The microsatellite PCR analysis of the *S. cerevisiae* strains, CLI 892 and CLI 889 showed that both strains are homozygous for the three alleles fingerprinted. The sizes of the alleles were as follows: SCPTSY7 (235 bp and 268 bp), SC8132X (215 bp and 209 bp), and YOR267C (416 bp and 451 bp), respectively. In the case of the strain isolated from spontaneous fermentations (S) allele sizes were as follows: SCPTSY7 (292 bp and 292 bp), SC8132X (212 bp and 310 bp), and YOR267C (308 bp and 389 bp). The comparison of these profiles with the five obtained by PFGE during the last phases of the fermentation of Malvar musts enabled that the number of patterns obtained and implantation rates during all fermentation processes in the different tanks were identical.

3.3. Principal Enological Parameters of the Fermentation Assays

Table 1 shows the means and standard deviations of the principal enological parameters, including organic acids of Malvar must fermentations carried out with the selected strains and uninoculated fermentations. Fermentative velocity (V_F) was higher in spontaneous fermentations, although the V_{50} was higher in the must inoculated with the yeast strain CLI 892. This finding is in agreement with the fact that those *S. cerevisiae* naturally present in must are better adapted to the fermentation conditions and environment than allochthonous inoculated yeasts [10]. It is interesting to point out that volatile acidity and acetic acid content were significantly lower in the uninoculated musts than the inoculated ones.

Regarding carboxylic acids, these compounds may contribute favorably to the organoleptic properties of young white wines. Our results showed that the total amount of these compounds was similar in wines from the CLI 892 strain and spontaneous fermentation, and higher than those obtained in wines from the CLI 889 strain. The excessive production of glycerol during wine fermentations for its positive sensory attributes gives rise to an increase in acetic acid concentration [26]. According to Erasmus *et al.* [27] it is, therefore, conceivable that different yeast strains experiencing the same fermentation conditions will respond by producing different concentrations of glycerol and acetic acid. The wine yeast strain isolated from the Malvar spontaneous fermentation (S), which produces low concentrations of acetic acid and conducts fermentations efficiently, seems to be a great candidate for the future production of high-quality Malvar wines in the Madrid winegrowing region, alone, or using mixed starter cultures.

Table 1. Enological parameters and organic acid concentrations of fermented wines (average and standard deviation (SD), Fisher's test (F, and significance (Sig) factors according to one-way ANOVA, test of comparison of means (Tukey): 2 and 11 degrees of freedom; the characters a, b, and c mean significant differences at $p \leqslant 0.05$. V_{50} = fermentation velocity consumption of 50% of the sugar content; V_F = fermentation velocity (% of daily sugar consumption).

Parameters	S	CLI 889	CLI 892	F	Sig
Alcoholic degree % (v/v)	12.50 ± 0.08 [a]	12.54 ± 0.06 [a]	12.70 ± 0.02 [b]	9.69	0.0132
Fermentative capacity	12.00 ± 0.07 [a]	12.61 ± 0.01 [c]	12.50 ± 0.01 [b]	186.53	0.0000
V_{50}	18.00 ± 0.04 [b]	15.45 ± 0.64 [a]	20.30 ± 0.2 [c]	128.70	0.0000
V_F	6.7 ± 0.3 [b]	5.40 ± 0.42 [a]	4.9 ± 0.7 [a]	10.27	0.0115
pH	3.81 ± 0.01 [b]	3.77 ± 0.01 [a]	3.79 ± 0.02 [b]	7.94	0.0206
Free SO_2 (mg/L)	6.0 ± 0.3 [a]	9.5 ± 0.71 [b]	10.0 ± 1.0 [b]	26.82	0.0010
Total SO_2 (mg/L)	15.0 ± 2.7	16.1 ± 1.56	14.0 ± 2.0	0.72	0.5230
Volatile acidity (g/L; acetic acid)	0.14 ± 0.01 [a]	0.23 ± 0.03 [b]	0.31 ± 0.04 [c]	25.04	0.0012
Titratable acidity (g/L; tartaric acid)	4.90 ± 0.07 [a]	5.14 ± 0.06 [b]	5.20 ± 0.07 [b]	16.93	0.0034
Citric acid (g/L)	0.31 ± 0.01 [b]	0.30 ± 0.00 [a,b]	0.29 ± 0.01 [a]	4.00	0.0787
Malic acid (g/L)	2.82 ± 0.05 [b]	2.58 ± 0.01 [a]	2.72 ± 0.09 [b]	27.44	0.0010
Lactic acid (g/L)	0.59 ± 0.06 [b]	0.55 ± 0.02 [b]	0.45 ± 0.04 [a]	8.36	0.0184
Acetic acid (g/L)	0.17 ± 0.01 [a]	0.32 ± 0.02 [b]	0.39 ± 0.03 [c]	69.14	0.0001
Succinic acid (g/L)	0.29 ± 0.01 [a,b]	0.28 ± 0.00 [a]	0.31 ± 0.01 [b]	5.55	0.0433
Reducing sugar (g/L)	1.5 ± 0.3	1.55 ± 0.07	1.3 ± 0.2	1.17	0.3730
Glycerol (g/L)	7.1 ± 1.6 [a,b]	7.85 ± 0.92 [b]	4.40 ± 1.5 [a]	5.13	0.0503
2,3-butanodiol (mg/L)	373.5 ± 67.5 [a]	386.8 ± 16.3 [a]	518.4 ± 91.2 [b]	7.86	0.0211

3.4. Aromatic Profile of Malvar Wines Fermented with Different Yeast Strains

Table 2 shows the average and standard deviations of the volatile compounds detected in the different fermentations. From all the volatile compounds identified, those presented at concentrations higher than their OTH (OAV higher than 1) are mainly considered as aroma-contributing compounds, and indicated in bold in Table 2.

Table 2. Data (Mean $\pm$ S.D.) of volatile composition related to the uninoculated fermentations (S) and inoculated fermentations with two locally-selected *S. cerevisiae* yeast strains (CLI 889 and CLI 892). Odor descriptors (ODE) and odor thresholds (OTH) described in the literature are included. Thresholds were calculated in a 10%–12% water/ethanol mixture. Odor activity values were also calculated. F = Fisher's test; Sig, one-way ANOVA analysis, test of comparison of means (Tukey): 2 and 11 degrees of freedom; the characters a, b, and c mean significant differences at $p \leqslant 0.05$.

Compound (mg/L)	S	CLI 889	CLI 892	F	Sig	ODE	OTH (mg/L)	OAV * S	OAV * CLI 889	OAV * CLI 892
Acetaldehyde	50.75 ± 8.85 [a]	69.61 ± 8.19 [b]	60.35 ± 10.30 [ab]	3.18	0.1142	Pleasant, fruity	0.0025 [2]	**20.30**	**27.84**	**24.14**
Acetoin	tr [a]	2.4 ± 1.03 [b]	1.27 ± 0.73 [ab]	8.14	0.0195	Flowery, wet	150.0 [1]	<0.1	<0.1	<0.1
Ethyl acetate	68.71 ± 1.01 [a]	69.28 ± 4.41 [a]	64.65 ± 10.66 [a]	0.43	0.6700	Fruit, solvent	12.26 [1]	**5.60**	**5.65**	**5.27**
1-Propanol	35.02 ± 1.83 [b]	36.63 ± 1.41 [b]	25.80 ± 2.58 [a]	25.63	0.0012	Alcohol, ripe fruit	306.0 [1]	0.11	0.11	<0.1
Isobutanol	37.39 ± 0.58 [b]	39.50 ± 1.53 [b]	23.52 ± 2.63 [a]	70.62	0.0001	Fusel, alcohol	40.00 [1]	0.93	0.98	0.58
Isoamylic alcohols	175.23 ± 1.74 [b]	182.20 ± 2.99 [c]	156.38 ± 3.25 [a]	71.47	0.0001	Bitter, harsh	30.00 [1]	**5.84**	**6.07**	**5.21**
Σ Higher major alcohols	247.65 ± 0.66 [b]	258.13 ± 5.63 [b]	205.71 ± 8.45 [a]	66.86	0.0001					
1-Hexanol	0.81 ± 0.01 [a]	0.95 ± 0.00 [a]	1.00 ± 0.06 [a]	0.42	0.6754	Green grass	8.00 [1]	0.1	0.11	0.12
2-Phenylethanol	10.79 ± 0.29 [b]	9.29 ± 0.85 [a]	9.99 ± 0.82 [ab]	3.46	0.100	Roses	14.00 [1]	0.77	0.66	0.71
Σ Higher minor alcohols	11.59 ± 0.27 [b]	10.22 ± 0.83 [a]	10.99 ± 0.80 [ab]	3.03	0.1233					

Table 2. *Cont.*

Compound (mg/L)	S	CLI 889	CLI 892	F	Sig	ODE	OTH (mg/L)	OAV* S	OAV* CLI 889	OAV* CLI 892
Isobutyl acetate	0.16 ± 0.00 b	0.18 ± 0.00 c	0.13 ± 0.01 a	37.75	0.0004	Sweet fruit	1.60 [1]	0.1	0.11	<0.1
Isoamyl acetate	7.66 ± 0.18 b	9.37 ± 0.54 c	6.19 ± 0.08 a	67.73	0.0001	Banana	0.030 [1]	**255.33**	**312.33**	**206.33**
Hexyl acetate	0.14 ± 0.01 b	0.19 ± 0.02 b	0.06 ± 0.05 a	11.07	0.0097	Fruity, green, pear	0.020 [3]	**7.00**	**9.50**	**3.00**
Phenylethyl acetate	0.56 ± 0.03 b	0.56 ± 0.07 b	0.38 ± 0.04 a	13.14	0.0064	Pleasant, flowery	0.250 [1]	**2.24**	**2.24**	**1.52**
Σ Higher alcohol acetates	8.38 ± 0.20 b	10.10 ± 0.62 c	6.70 ± 0.04 a	61.06	0.0001					
Ethyl butyrate	0.33 ± 0.01 ab	0.39 ± 0.00 b	0.27 ± 0.06 a	10.24	0.0116	Acid fruit	0.020 [1]	**16.5**	**19.50**	**13.5**
Ethyl hexanoate	0.60 ± 0.05 b	0.52 ± 0.07 ab	0.39 ± 0.13 a	4.16	0.0735	Green apple	0.014 [1]	**42.86**	**37.14**	**27.86**
Ethyl octanoate	1.01 ± 0.25 a	1.10 ± 0.11 a	0.73 ± 0.42 a	1.34	0.3314	Sweet, soap	0.005 [1]	**202.00**	**220.00**	**146.00**
Ethyl decanoate	0.08 ± 0.06 b	0.23 ± 0.02 c	tr a	28.09	0.0009	Pleasant, soap	0.200 [1]	0.40	**1.15**	<0.1
Σ Fatty acid esters	2.01 ± 0.37 a	2.22 ± 0.19 a	1.40 ± 0.0.58 a	3.21	0.1127					
Ethyl lactate	tr a	2.24 ± 1.01 b	tr a	14.57	0.005	Lactic	0.157 [2]	<0.1	**14.26**	<0.1
Diethyl succinate	0.03 ± 0.01 a	0.09 ± 0.04 b	0.05 ± 0.01 ab	4.67	0.0599	Apple, fruity	0.20 [2]	0.15	0.45	0.25
Isobutyric acid	3.01 ± 0.02 a	3.02 ± 0.03 a	2.91 ± 0.01 a	0.94	0.4429	acid, fatty	0.230 [1]	**13.09**	**13.13**	**12.65**
Butyric acid	1.90 ± 0.04 a	1.93 ± 0.21 a	1.31 ± 1.14 a	0.81	0.4876	Cheese	0.173 [1]	**10.98**	**11.15**	**7.57**
Isovaleric acid	0.08 ± 0.06 a	0.26 ± 0.07 b	0.02 ± 0.01 a	14.60	0.005	Blue cheese	0.033 [1]	**2.42**	**7.87**	0.60
Σ SCFA	6.67 ± 0.07 a	6.65 ± 0.15 a	6.08 ± 1.06 a	0.88	0.4640					
Hexanoic acid	4.97 ± 0.23 a	4.29 ± 0.36 a	4.54 ± 0.11 ab	5.27	0.0477	Cheese	0.420 [1]	**11.83**	**10.21**	**10.80**
Octanoic acid	6.11 ± 0.20 c	5.45 ± 0.50 b	5.07 ± 0.20 a	17.91	0.0030	Rancid, harsh	0.500 [1]	**12.22**	**10.90**	**10.14**
Decanoic acid	3.70 ± 0.77 a	3.68 ± 0.32 a	2.85 ± 0.21 a	2.89	0.1319	Fatty	1.00 [1]	**3.70**	**3.68**	**2.85**
Σ MCFA	14.78 ± 1.20 b	13.39 ± 1.15 ab	12.46 ± 0.35 a	4.23	0.0713					
4-Vinylguaiacol	0.47 ± 0.18 ab	0.53 ± 0.04 b	0.29 ± 0.09 a	3.33	0.1062	Pleasant, phenolic	1.10 [1]	0.42	0.48	0.26

tr = traces; 1 = thresholds from Gil *et al.* [20]; 2 = thresholds from Duarte *et al.* [28]; 3 = threshold from Falqué *et al.* [29]; SCFA Small Chain Fatty Acids; MCFA Medium Chain Fatty Acids; * in bold, compounds with OAV > 1.

In all cases, achieved acetaldehyde levels were higher than the threshold proposed by Duarte *et al.* [28]. Among these values the highest values corresponds to the inoculated musts, while in the spontaneous fermentations the values were slightly lower (Table 2). Only free acetaldehyde has flavor relevance; at low levels it provides fruity flavors, while high concentrations (>200 mg/L) contribute "flatness" in wines [20]. All of the studied Malvar wines can be treated as correct due to their mean content being within the range previously studied in other non-oxidized white wines [30]. High concentration of acetoin is mostly related to non-*Saccharomyces* fermentations [1,8] but in our study, high concentrations were only detected in the inoculated fermentations. On the other hand, ethyl acetate concentrations showed higher values than the threshold proposed (OAVs, from 5.2 to 5.6) over all Malvar wines (Table 2). Ethyl acetate may confer with pleasant and fruity fragrances to the global wine aroma at concentrations lower than 150 mg/L.

Quantitatively, the largest group of volatile compounds present in wines were higher alcohols. Many of these compounds are strongly correlated with dislikable aromas in wines, but even though significant differences were found among the uninoculated and inoculated wines, the values obtained in wines fermented by the CLI 892 strain were significantly lower than the other two wines (Table 2). The amount of higher alcohols ranged from 205.71 ± 8.45 to 258.13 ± 5.63 mg/L, this variation principally is due to the isoamyl alcohol produced, which represent in all cases close to 60% of the total higher alcohols (Table 2). Discrete concentrations of fusel alcohols contribute to the wine's aromatic complexity. With respect to the higher minor alcohols, this group of volatiles had no odor activity contribution.

In terms of the number of components analyzed, esters represent the largest group (Table 2). Higher alcohol acetates are an important group of fermentative aromas, which are normally linked to fruity descriptors, but in a wine can significantly modify the global aroma. The fruity character associated to the aroma of Madrid white wines is mainly related to fruit notes such as banana, green apple, or pear [20,31,32], produced by the acetates of higher alcohols and fatty acids.

Significant differences were detected among the different fermentations for isoamyl acetate, hexyl acetate, and phenyl ethyl acetate (Table 2). The major ester in the Malvar wines analyzed was isoamyl acetate which was present in high concentrations and above of the OTH and OAV in all fermentations. Ethyl esters of fatty acids showed important variations in the concentrations, but in the majority of the cases, they showed higher values than the OTH and OAVs values proposed (Table 2). The real contribution of ethyl lactate and diethyl succinate to the white wines of the "D.O. Vinos de Madrid" has been previously described as insignificant [20,32]. Interestingly, Malvar wines fermented by the yeast strain CLI 889 showed a high concentration of ethyl lactate. It is now widely understood that ester concentration is conditioned by several factors, such as yeast strain, fermentation temperature, aeration degree and sugar content [33–35]. In our experimental conditions, in spite of the use of the same must under the same conditions, the different amounts of esters found can be attributed to the yeast strain used.

Fatty acids have been described with cheese, harsh, fatty, and rancid notes [20]. All compounds in the different fermentations reached the OTH and OAVS values proposed (Table 2), with the exception of the isovaleric acid in the fermentations carried out by CLI 892. Supposing that the presence of fatty acids is frequently associated to off flavors, they play an important role in the aromatic equilibrium in wines because they are antagonistic to the hydrolysis of the analogous esters [20].

High concentrations of vinyl phenols can be responsible for strong pharmaceutical odors in white wines [18,36], but at low or moderate concentrations they could be linked with grassy, herbaceous, or pleasant spicy aromas. Accordingly, Grando et al. [37] pointed out that 4-vinylguaiacol was the principal responsible compound for the spicy aroma of Gewürztraminer's wines. In our study, 4-vinylguaiacol was detected among the uninoculated and inoculated white wines. Statistically significant differences were observed when using Malvar juices fermented spontaneously or inoculated with the selected yeast strains.

PCA was performed to disclose the compounds that differentiated best among the uninoculated and inoculated Malvar wines (Figure 3). Hence, only yeast volatile fermentation constituents enabled a suitable discrimination between inoculated and spontaneous fermentation wines.

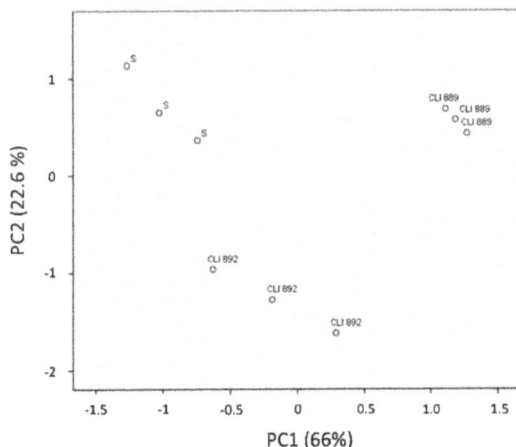

Figure 3. Principal components analysis of volatile composition data. Projection of the Malvar wines fermented spontaneously (S), and with CLI 892 and CLI 889 *S. cerevisiae* strains in the dials formed by the PC1 (22.6%) and PC2 (66%).

The PCA explained the 88.63% of the total variance. Musts fermented with the strain CLI 889 shaped a clear group, which was associated with the esters ethyl octanoate, isoamyl acetate, and ethyl lactate, as long as fermentation inoculated with CLI 892 was located in a different dial of the PCA plot and most associated with those compounds contained in the PC2 (ethyl hexanoate, 4-vinylguaiacol, ethyl butyrate, and ethyl acetate). The uninoculated wines showed a cluster mostly associated with those compounds of the PC1, but none was nearly correlated with any of the inoculated fermentation wines (data not shown). Thus, taking into account that fermentations were conducted in the same must, if the yeast starter dominates on native yeast population, the wine will exhibit singular aroma and sensory profiles of the each yeast starter involved [14]. Confirming this, PCA analysis and data of Table 2 showed that by comparison with inoculated wines with the yeast strains CLI 889 and CLI 892, uninoculated fermentation wines showed a high variability in the composition of volatile compounds that also contributes to wine aroma.

3.5. Sensory Analysis of Wines

After the first sensory analysis using a penalizing system on all of the wines, attributes related to the appearance, taste intensity, taste quality, and harmony were not statistically different. However, differences ($p < 0.05$) were found within the different Malvar wines in terms of their aroma intensity, and quality. The ratings obtained were; uninoculated 15.73 ± 3.20, inoculated with the yeast strain CLI 889 13.30 ± 1.92, and 16.40 ± 4.55 using the strain CLI 892. Thus, judges considered that wine made by inoculation of CLI 889 seem to have the better quality. On the other hand, the sensory analysis using the different selected descriptors reached that some descriptors were statistically influenced by the yeast strain (Figure 4).

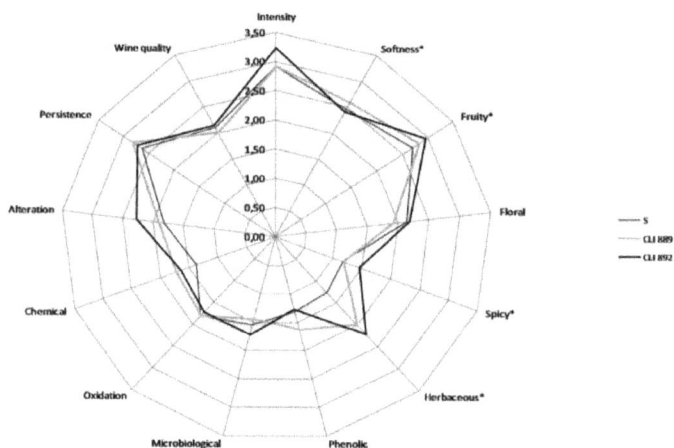

Figure 4. Polar coordinate (cobweb) graph of mean sensory scores rating of "alteration", "chemical", "herbaceous", "intensity", "floral", "fruity", "microbiological", "oxidation", "persistence", "phenolic", "softness", "spicy", and "wine quality" for wines made with the uninoculated *S. cerevisiae* (S) and locally-selected CLI 889 and CLI 892 yeast strains. In sensorial variables indicated with an asterisk (*) a difference between some trials is verified for $p \leqslant 0.05$.

The greatest softness of the spontaneously fermented wines was deserved to a lower acidity, and apparently to a large production of glycerol and as reported by Feuillat [38]. Glycerol is sweet, however its contribution to the sweet taste or palatableness of a wine remains unclear. The threshold concentration proposed by Noble and Bursick [39] is considered high (5.2 g/L) as quoted by Ugliano and Henschke [40], and the high acidity in wines is wont to interact with the remaining sweetness, so the limit to which glycerol contributes to sweetness or mouthfeel characters is still unrevealed. Nevertheless, is in agreement with Scacco *et al.* [41], at the concentrations obtained in this work (Table 1), glycerol contributes positively to softness and viscosity of the wines. The differences concerning fruitiness cannot be considered to be related to the amount of wine flavor compounds. Substantially, wines fermented by the strain CLI 892 was characterized by a considerable content in higher alcohol acetates (10.10 ± 0.62 mg/L) with banana, pear, and herbaceous notes, followed by spontaneous fermentation with 8.38 ± 0.62 mg/L (Table 2). These compounds are usually linked to fruity descriptors; hence, it is possible to assume that the fruity descriptor in wines made with CLI 889 and CLI 892 were swayed by the lower amount of substances with a masking effect on fruity, than in the spontaneous fermentations, as reported by Campo *et al.* [42]. The spiciness was lower in wines fermented by strain CLI 892, apparently because this strain produced a lower quantity of some compounds able to influence this descriptor than the other strains, such as 4-vinylguaiacol. Sensory analysis showed that the best wines were those fermented by CLI 889, which had great gustatory and aromatic characteristics, usual in high-quality white wines. These results could be associated to a reasonable production of several fruity esters, such as isoamyl acetate, and hexyl acetate, which highlight the organoleptic properties and regional characteristic of Malvar white wines.

4. Conclusions

In the context of what is covered in this work, winemakers of the winegrowing regions where Malvar is cultivated are set to benefit from this recent development, because they will have improved options to elaborate wine. Uninoculated and locally-selected yeast strains clearly transform the fermentation and have influence over the volatile profile of Malvar white wines. In spite of wines fermented spontaneously and yeast strains CLI 889 and 892 being considered accurate by judges, the wine fermented with the strain CLI 889 was qualified as the best. The most salient chemical attributes were its high concentrations of isoamyl acetate and hexyl acetate and the high acidity, which have been detected to be key constituents in setting the fruity and fresh character of Malvar wines. Thus, this autochthonous yeast strain has been deposited at the Spanish Type Collection Culture (CECT) with the accession number CECT 13145. In addition, a new low acetic acid-producing *S. cerevisiae* yeast strain isolated from the spontaneous fermentations was characterized. Other features of its contribution to the general characteristics of these white wines are now being investigated in order to enrich the Malvar wine quality.

Supplementary Materials: Supplementary materials can be found at http://www.mdpi.com/2311-5637/2/1/7/s1.

Acknowledgments: We are grateful to Matthew Tyndale-Tozer for revising the manuscript. This work was financed by the Instituto Nacional de Investigación y Tecnología Agraria y Alimentaria (INIA- RM 2006-00012-00-00), by the Instituto Madrileño de Investigación y Desarrollo Rural Agrario y Alimentario (IMIDRA-FP07-AL2-Lev), and by the European Union's Seventh Framework Programme via the Marie Curie Action, "Co-funding of Regional, National and International Programs" to stimulate research activities without mobility restrictions, co-financed by the "Junta de Andalucía" and the European Commission under grant agreement no 291780.

Author Contributions: All authors contributed to the conception and design of experiments; Mar Gil-Díaz and Margarita García performed the experiments, Gustavo Cordero-Bueso performed the sensory analysis; all authors analyzed the data; Gustavo Cordero-Bueso and Teresa Arroyo contributed to writing the paper.

Conflicts of Interest: The authors declare no conflict of interest. The founding sponsors had no role in the design of the study; in the collection, analyses, or interpretation of data; in the writing of the manuscript, and in the decision to publish the results.

References

1. Cordero-Bueso, G.; Esteve-Zarzoso, B.; Cabellos, J.M.; Gil-Díaz, M.; Arroyo, T. Biotechnological potential of non-*Saccharomyces* yeasts isolated during spontaneous fermentations of Malvar (*Vitis viniferacv* L). *Eur. Food Res. Technol.* **2013**, *236*, 193–207. [CrossRef]

2. Laffort, J.F.; Romat, H.; Darriet, P. Les levuresetl'expressionaromatique des vinsblancs. *Rev. Oenol.* **1989**, *53*, 9–12.

3. Günata, Y.Z.; Dugelay, I.; Sapis, J.C.; Baumes, R.; Bayonove, C. Role of enzymes in the use of the flavor potential from grape glycosides in winemaking. In *Progress in Flavor Precursor Studies*; Schreier, P., Winter-Halter, P., Eds.; Allured Publishing Corporation: Carol Stream, IL, USA, 1993; Volume 3, pp. 219–234.

4. Swiegers, J.H.; Bartowsky, E.J.; Henschke, P.A.; Pretorius, I.S. Yeast and bacteria modulation of wine aroma and flavor. *Aust. J. Grape Wine Res.* **2005**, *11*, 139–173. [CrossRef]

5. Swiegers, J.H.; Pretorius, I.S. Yeast modulation of wine flavor. *Adv. Appl. Microbiol.* **2007**, *57*, 131–175.

6. Carrascosa, A.V.; Bartolomé, B.; Robredo, A.; León, A.; Cebollero, E.; Juega, M.; Nuñez, Y.P.; Martínez, M.C.; Martínez-Rodríguez, A.J. Influence of locally-selected yeast on the chemical and sensorial properties of Albariño White wines. *LWT Food Sci. Technol.* **2012**, *46*, 319–325. [CrossRef]

7. Varela, C.; Siebert, T.; Cozzolino, D.; Rose, L.; McLean, H.; Henschke, P. Discovering a chemical basis for differentiating wines made by fermentation with wild indigineous and inoculated yeasts; role of yeast volatile compounds. *Aust. J. Grape Wine Res.* **2009**, *15*, 238–248. [CrossRef]

8. Vigentini, I.; Fabrizio, V.; Faccincani, M.; Picozzi, C.; Comasio, A.; Foschino, R. Dynamics of *Saccharomyces cerevisiae* populations in controlled and spontaneous fermentations for Franciacorta D.O.C.G. base wine production. *Ann. Microbiol.* **2013**, *64*, 639–651. [CrossRef]

9. Capece, A.; Pietrafesa, R.; Romano, P. Experimental approach for target selection of wild wine yeasts from spontaneous fermentation of "Inzolia" grapes. *World J. Microbiol. Biotechnol.* **2011**, *27*, 2775–2783. [CrossRef]

10. Csoma, H.; Zakany, N.; Capece, A.; Romano, P.; Sipiczki, M. Biological diversity of *Saccharomyces* yeasts of spontaneously fermenting wines in four wine regions: Comparative genotypic and phenotypic analysis. *Int. J. Food Microbiol.* **2010**, *140*, 239–248. [CrossRef] [PubMed]

11. Esteve-Zarzoso, B.; Gostincar, A.; Bobet, R.; Uruburu, F.; Querol, A. Selection and molecular characterization of wine yeasts isolated from the "El Penedes" area Spain. *Food Microbiol.* **2000**, *17*, 553–562. [CrossRef]

12. Orlic, S.; Redzepovic, S.; Jeromel, A.; Herjavec, S.; Iacumin, L. Influence of indigenous *Saccharomyces paradoxus* strains on Chardonnay wine fermentation aroma. *Int. J. Food Sci. Technol.* **2007**, *42*, 95–101. [CrossRef]

13. Fleet, G.H. Wine yeast for the future. *FEMS Yeast Res.* **2008**, *8*, 979–995. [CrossRef] [PubMed]

14. Rodríguez, M.E.; Infante, J.J.; Molina, M.; Domínguez, M.; Rebordinos, L.; Cantoral, J.M. Genomic characterization and selection of wine yeast to conduct industrial fermentations of a white wine produced in a SW Spain winery. *J. Appl. Microbiol.* **2010**, *108*, 1292–1302. [CrossRef] [PubMed]

15. Suzzi, G.; Arfelli, G.; Schirone, M.; Corsetti, A.; Perpetuini, G.; Tofalo, R. Effect of grape indigenous *Saccharomyces cerevisiae* strains on Montepulciano d'Abruzzo red wine quality. *Food Res. Int.* **2012**, *46*, 22–29. [CrossRef]

16. Arroyo, T. Estudio de la influencia de diferentes tratamientos enológicos en la evolución de la microbiota y en la calidad de los vinos elaborados con la variedad "Airén", en la D.O. "Vinos de Madrid". Ph.D. Thesis, University of Alcalá, Alcalá de Henares, Madrid, Spain, 2000.

17. Cordero-Bueso, G.; Arroyo, T.; Serrano, A.; Tello, J.; Aporta, I.; Vélez, M.D.; Valero, E. Influence of the farming system and vine variety on yeast communities associated with grape-berries. *Int. J. Food Microbiol.* **2011**, *45*, 132–139. [CrossRef] [PubMed]

18. Esteve-Zarzoso, B.; Belloch, C.; Uruburu, F.; Querol, A. Identification of yeast by RFLP analysis of the 5.8 rRNA and the two ribosomal internal transcribed spacers. *Int. J. Syst. Bacteriol.* **1999**, *49*, 329–337. [CrossRef] [PubMed]

19. Vaudano, E.; Garcia-Moruno, E. Discrimination of *Saccharomyces cerevisiae* wine strains using microsatellite multiplex PCR and band pattern analysis. *Food Microbiol.* **2008**, *25*, 56–64. [CrossRef] [PubMed]

20. Gil, M.; Cabellos, J.M.; Arroyo, T.; Prodanov, M. Characterization of the volatile fraction of young wines from the Denomination of Origin "Vinos de Madrid" Spain. *Anal. Chim. Acta* **2006**, *563*, 145–153. [CrossRef]

21. International Organization for Standardization. *ISO 11035. Sensory Analysis—Identification and Selection of Descriptors for Establishing a Sensory Profile by Multidimensional Approach*; ISO: Geneva, Switzerland, 1994.

22. International Organization for Standardization. *ISO 8589. Sensory Analysis—General Guidance for the Design of Test Room*; ISO: Geneva, Switzerland, 2007.

23. International Organization for Standardization. *ISO 8586-1. Sensory Analysis—General Guidance for the Selection, Training, and Monitoring of Assessors. Part 1: Selected Assessors*; ISO: Geneva, Switzerland, 1993.

24. Lozano, J.; Santos, J.P.; Arroyo, T.; Aznar, M.; Cabellos, J.M.; Gil, M.; Horrillo, M.C. Correlating e-nose responses to wine sensorial descriptors and gas Chromatography—Mass spectrometry profiles using partial least squares regression analysis. *Sens. Actuators B* **2007**, *127*, 267–276. [CrossRef]

25. Vilanova, M.; Masneuf-Pomarède, I.; Dubourdieu, D. Influence of *Saccharomyces cerevisiae* strains on general composition and sensorial properties of white wines made from Vitis vinifera cv. Albariño. *Food Technol. Biotechnol.* **2005**, *43*, 79–83.

26. Remize, F.; Roustan, J.L.; Sablayrolles, J.M.; Barre, P.; Dequin, S. Glycerol overproduction by engineered *Saccharomyces cerevisiae* wine yeast strains leads to substantial changes in byproduct formation and to a stimulation of fermentation rate in stationary phase. *Appl. Environ. Microbiol.* **1999**, *65*, 143–149. [PubMed]

27. Erasmus, D.J.; Cliff, M.; van Vuuren, H.J.J. Impact of yeast strain on the production of acetic acid, glycerol, and the sensory attributes of icewine. *Am. J. Enol. Viticult.* **2004**, *55*, 371–378.

28. Duarte, F.W.; Dias, R.D.; Oliveira, J.M.; Vilanova, M.; Teixeira, J.A.; Almeida e Silva, J.B.; Schwan, R.F. Raspberry *Rubusidaeus* L. wine: Yeast selection, sensory evaluation and instrumental analysis of volatile and other compounds. *Food Res. Int.* **2010**, *43*, 2303–2314. [CrossRef]

29. Falqué, E.; Fernández, E.; Dubourdieu, D. Differentation of white wines by their aromatic index. *Talanta* **2001**, *54*, 271–281. [CrossRef]

30. Escudero, A.; Asensio, E.; Cacho, J.; Ferreira, V. Sensory and chemical changes of young white wines stored under oxygen. An assessment of the role played by aldehydes and some other important odorants. *Food Chem.* **2002**, *77*, 325–331. [CrossRef]

31. Arroyo, T.; Lozano, J.; Cabellos, J.M.; Gil-Díaz, M.; Santos, J.P.; Horrillos, M.C. Evaluation of wine aromatic compounds by a sensory human panel and an electronic nose. *J. Agric. Food Chem.* **2009**, *57*, 11449–11582. [CrossRef] [PubMed]

32. Santos, J.P.; Arroyo, T.; Aleixandre, M.; Lozano, J.; Sayago, I.; García, M.; Fernández, M.J.; Arés, L.; Gutiérrez, J.; Cabellos, J.M.; *et al.* A comparative study of sensor array and GC-MS: application to Madrid wines characterization. *Sens. Actuators B* **2004**, *102*, 299–307. [CrossRef]

33. Hernández-Orte, P.; Cersosimo, M.; Loscos, N.; Cacho, J.; García-Moruno, E.; Ferreira, V. The development of varietal aroma from non-floral grapes by yeast of different genera. *Food Chem.* **2008**, *107*, 1064–1077. [CrossRef]

34. Rojas, V.; Gil, J.V.; Piñaga, F.; Manzanares, P. Acetate ester formation in wine by mixed cultures in laboratory fermentations. *Int. J. Food Microbiol.* **2003**, *86*, 181–188. [CrossRef]

35. Valero, E.; Moyano, L.; Millán, M.C.; Medina, M.; Ortega, J.M. Higher alcohols and esters production by *Saccharomyces cerevisiae*. Influence of the initial oxygenation of the grape must. *Food Chem.* **2002**, *78*, 57–61. [CrossRef]

36. Gómez-Míguez, M.J.; Cacho, J.F.; Ferreira, V.; Vicario, I.M.; Heredia, F.J. Volatile components of Zalema white wines. *Food Chem.* **2007**, *100*, 1464–1473. [CrossRef]

37. Grando, M.S.; Versini, G.; Nicolini, G.; Mattivi, F. Selective use of wine yeast strains having different volatile phenols production. *Vitis* **1993**, *32*, 43–50.

38. Feuillat, M. Yeast macromolecules: Origin, composition, and enological interest. *Am. J. Enol. Vitic.* **2003**, *54*, 211–213.

39. Noble, A.C.; Bursick, G.F. The contribution of glycerol to perceived viscosity and sweetness in white wine. *Am. J. Enol. Vitic.* **1984**, *35*, 110–112.

40. Ugliano, M.; Henschke, P.A. Yeasts and wine flavour. In *Wine Chemistry and Biochemistry*; Moreno-Arribas, M.V., Polo, M.C., Eds.; Springer Verlag: New York, NY, USA, 2009; pp. 313–392.

41. Scacco, A.; Oliva, D.; Di Maio, S.; Polizzotto, G.; Genna, G.; Tripodi, G.; Lanza, C.M.; Verzera, A. Indigenous *Saccharomyces cerevisiae* strains and their influence on the quality of Cataratto, Inzolia and Grillo white wines. *Food Res. Int.* **2012**, *46*, 1–9. [CrossRef]

42. Campo, E.; Ferreira, V.; Escudero, A.; Marqués, J.C.; Cacho, J. Quantitative gas chromatography-olfactometry and chemical quantitative study of the aroma of four Madeira wines. *Anal. Chim. Acta* **2006**, *563*, 180–187. [CrossRef]

fermentation

MDPI

Article

The Interaction of Two *Saccharomyces cerevisiae* Strains Affects Fermentation-Derived Compounds in Wine

Frida S. Gustafsson [1], Vladimir Jiranek [2], Marissa Neuner [1], Chrystal M. Scholl [1], Sydney C. Morgan [1] and Daniel M. Durall [1,*]

[1] The University of British Columbia (UBC), Okanagan, Biology Department, 1177 Research Rd., Kelowna, BC V1V 1V7, Canada; fridasofiegustafsson@gmail.com (F.S.G.); marissa.neuner@ubc.ca (M.N.); chrystalmarie.scholl@gmail.com (C.M.S.); sydney-morgan@hotmail.com (S.C.M.)
[2] Department of Wine and Food Science, The University of Adelaide, PMB1, Glen Osmond SA 5064, Australia; vladimir.jiranek@adelaide.edu.au
*   Correspondence: daniel.durall@ubc.ca; Tel.: +1-250-807-8759

Academic Editor: Ronnie G. Willaert
Received: 31 December 2015; Accepted: 18 March 2016; Published: 30 March 2016

Abstract: Previous winery-based studies showed the strains Lalvin® RC212 (RC212) and Lalvin® ICV-D254 (D254), when present together during fermentation, contributed to >80% relative abundance of the *Saccharomyces cerevisiae* population in inoculated and spontaneous fermentations. In these studies, D254 appeared to out-compete RC212, even when RC212 was used as the inoculant. In the present study, under controlled conditions, we tested the hypotheses that D254 would out-compete RC212 during fermentation and have a greater impact on key fermentation-derived chemicals. The experiment consisted of four fermentation treatments, each conducted in triplicate: a pure culture control of RC212; a pure culture control of D254; a 1:1 co-inoculation ratio of RC212:D254; and a 4:1 co-inoculation ratio of RC212:D254. Strain abundance was monitored at four stages. Inoculation ratios remained the same throughout fermentation, indicating an absence of competitive exclusion by either strain. The chemical profile of the 1:1 treatment closely resembled pure D254 fermentations, suggesting D254, under laboratory conditions, had a greater influence on the selected sensory compounds than did RC212. Nevertheless, the chemical profile of the 4:1 treatment, in which RC212 dominated, resembled that of pure RC212 fermentations. Our results support the idea that co-inoculation of strains creates a new chemical profile not seen in the pure cultures. These findings may have implications for winemakers looking to control wine aroma and flavor profiles through strain selection.

Keywords: *Saccharomyces cerevisiae*; strain interaction; fermentation-derived compounds

1. Introduction

In spontaneous fermentations conducted at commercial wineries, it is common to find more than one *Saccharomyces cerevisiae* strain fermenting the wine must [1]; however, multiple strains have also been detected even in inoculated fermentations [2,3]. It is well documented that different wine strains of *S. cerevisiae* affect flavor and aroma properties differently [1]. Although the sensory influence of co-inoculation between non-*Saccharomyces* and a single *S. cerevisiae* strain has been widely studied [4–8], fewer studies have reported on the co-inoculation of multiple *S. cerevisiae* strains [9–13]. The commercial active dry yeast (ADY) strains, Lalvin® Bourgorouge RC212 (RC212) and Lalvin® ICV-D254 (D254), are frequently used to ferment Pinot Noir and Chardonnay musts, respectively. Together, they have been found to dominate operational fermentations, with an overall relative abundance of >80% in both inoculated (where RC212 was used as the sole inoculum and D254 entered as a contaminant) and spontaneous Pinot Noir fermentations [14,15]. Furthermore, D254 was the dominant strain at the end of these fermentations, even when tanks were inoculated with RC212 [3]. These findings suggest, when observing their dynamics during operationally conducted fermentations, that D254 out-competes RC212. Originally, the strain RC212 was selected by the Burgundy Wine Board (BIVB) to extract and protect the polyphenols of Pinot Noir. In the information supplied by the manufacturer, it is claimed that wines fermented by RC212 have good structure with fruity and spicy characteristics (Lallemand Inc., Montreal, QC, Canada). The strain D254 is commonly used in both red and white wines. Red wines fermented with D254 contribute to high fore-mouth volume, smooth tannins, intense fruits and a slightly spicy finish (Lallemand Inc., Montreal, QC, Canada). Nevertheless, there is a lack of information on the sensorial attributes when these two strains co-exist during fermentation. Given that there are many factors that can affect the interactions of these two strains under operational conditions, it is important to determine how these two strains interact and affect key fermentation-derived chemicals under controlled conditions.

The formation of aroma and flavor compounds is dependent on the nutrient availability, the physicochemical properties of the fermentation, and the yeast strains present, especially *S. cerevisiae* strains. Higher alcohols and esters are usually yeast-derived and can greatly contribute to the aroma and flavor profile of the wine [16]. Many of these flavor compounds are derivatives of amino acids, and it has been shown that amino acid uptake by yeasts is strain-dependent [11,17]. Other wine aroma and flavor compounds include pyrazines, terpenes, lactones, sulfur-containing compounds, phenols, organic acids, and aldehydes, which are usually not strain-dependent. The concentration of these other compounds is strongly influenced by varietal, grape ripeness, non-*Saccharomyces* organisms, aging, and winemaking practices [16,18]. Several studies have concluded that different strains of *S. cerevisiae* produce strain-specific metabolites [19,20]. For example, higher alcohols and esters can differ with varying dominance of two or more strains [11,19,20]. At low concentrations, higher alcohols contribute to increased aroma complexity, but at high concentrations (>300 mg/L), their presence can be undesirable [21,22]. At low concentrations (<100 mg/L), ethyl esters, such as ethyl acetate, often contribute fruity aromas, but at high concentrations they can produce undesirable solvent-like aromas and flavors [16,23]. In the present study, we targeted only compounds that are known to be fermentation-derived and are integral to aroma and flavor development.

Knowledge of the competitive interaction between different *S. cerevisiae* strains and its effect on aroma and flavor compounds will guide winemakers in choosing commercial yeasts, because final wine composition may be enhanced with the use of the most suitable combination of yeast strains [11]. In addition, we are not aware of any competition or metabolomic studies that have conducted co-fermentations with RC212 and D254 strains in grape must. For our study, competition between two strains, which ultimately results in competitive exclusion, is defined at the end of a co-inoculated fermentation, where one strain has a greater relative abundance than it did when it was inoculated.

The aim of this study was to generate and test hypotheses that were based on observations from operational settings and from the literature. We tested, under controlled conditions, the hypotheses that: (1) D254 will out-compete RC212 when inoculated as a 1:1 or as a 4:1 RC212:D254 ratio; (2) D254 will have a greater impact than RC212 on key fermentation-derived chemicals when the inoculation abundances of the two strains are equal; and (3) D254 will have a greater impact than RC212 on key fermentation-derived chemicals when the inoculation is administered in a 4:1 RC212:D254 ratio. Our results indicate that no competitive exclusion occurred in the co-inoculated treatments, but rather the inoculated ratios remained constant throughout fermentation. Furthermore, we found that the chemical profile of the 1:1 RC212:D254 treatment closely resembled the chemical profile of the pure D254 fermentations, but the 4:1 RC212:D254 treatment more closely resembled the chemical profile of the pure RC212 fermentations. We conclude that although D254 does not appear to competitively exclude RC212 under controlled conditions, it has a relatively larger impact on the sensory profile of the resulting wines than RC212.

2. Materials and Methods

2.1. Experimental Design

The experiment consisted of four fermentation treatments: a pure culture control of RC212; a pure culture control of D254; a 1:1 co-inoculation ratio of RC212:D254; and a 4:1 co-inoculation ratio of RC212:D254. Each treatment was replicated using three separate fermentation flasks for a total of 12 flasks, with each flask containing 100 mL Pinot Noir juice. Each flask was sampled for strain abundance at the start (180 g/L sugar, 0 h), early (83–102 g/L sugar, 24 h), mid (64–73 g/L sugar, 32 h), and end stages (<2 g/L sugar, 97 h) of the 100 h fermentation. Samples for chemical analysis were taken only at the end stage of fermentation. The co-inoculation treatments represented one situation where the two strains were inoculated in equal abundance (1:1 ratio) and another where RC212 was inoculated at a higher proportion than D254 (4:1 ratio); these two co-inoculation treatments, along with their pure-culture controls, allowed us to adequately test all of our hypotheses.

2.2. Juice Preperation

Pinot Noir juice was obtained from WineExpert™ (Port Coquitlam, BC, Canada). The juice was prepared by centrifugation for 45 min at 3500× g and was subsequently filtered through a series of filters, which had a decreasing pore size: 2.7 µm glass fiber filter (GF), 1 µm GF, 0.45 µm mixed cellulose ester membrane filter (MCE), and 0.22 µm MCE and polyvinylidene difluoride filter (PVDF). The filtered juice was adjusted to 180 g/L sugar with sterile Milli-Q water and stored at -20 °C until it was needed for the experiment. We selected this concentration because it was within the typical range (180–220 g/L) at which grape juice fermentation commences [24]. The filtered juice, following the adjustment to 180 g/L sugar, had a pH of 3.8 and its sterility was confirmed by plating 0.1 mL onto yeast extract peptone dextrose (YEPD) media and observing an absence of colonies after 4 days of incubation at 28 °C. The adjusted filtered juice (>2 L) was used as the source to make the RC212 and D254 inoculated solutions, described in the section below.

2.3. Inoculation and Fermentations

For both strains, (~10 mg) ADY inoculum was rehydrated in 25 mL liquid YEPD media and was shaken for eight hours (120 rpm) at 28 °C. Yeast abundance (cells/mL) in the rehydrated suspension was counted using a hemocytometer. Rehydrated yeasts were added in a quantity of 1×10^6 cells/mL to 100 mL diluted Pinot Noir grape juice (1:1 juice:sterile Milli-Q H_2O) for each strain. Once yeast cell count was determined in each solution, the RC212 and D254 solutions were added separately to 1.2 L and 700 mL of the filtered Pinot Noir juice, respectively, to produce a concentration of 5×10^6 cells/mL. The resulting master mixes of each strain were combined in the appropriate ratios to obtain 300 mL of each co-inoculation treatment. Subsequently, for each co-inoculation treatment, the resulting solution was divided into three independent flasks (each containing 100 mL juice). The pure-culture controls were treated the same way; however, the RC212 and D254 solutions were not mixed. For all treatments, the final inoculation concentration was 5×10^6 cells/mL.

Fermentations (100 mL per flask) were conducted in 250 mL fermentation flasks, which contained sampling ports and air-locks. The flasks were shaken (120 rpm) at 28 °C until the end of the fermentation. To monitor the progression of fermentation and to identify strains, 0.5 mL samples were collected aseptically at the start, early, mid, and end stages of fermentation. At the end of fermentation, all wines contained <2 g/L residual sugar, as indicated with a D-Glucose/D-Fructose sugar assay kit (Megazyme, Bray, Ireland). The wine was clarified by centrifugation (1200× g; 2 min) and filtered (0.45 µm) at the end of fermentation. At the end stage, 40 mL were transferred to glass vials and stored at -80 °C until chemical analysis was performed.

2.4. Yeast Strain Identification

Wine must samples from each stage were plated on YEPD agar and incubated at 28 °C for 48 h. Twenty colonies from each plate (960 colonies total) were randomly chosen for DNA analysis. Extraction and amplification of the DNA followed the methods of Lange *et al.* [15], except that amplification of the isolates was performed with primer sets for the microsatellite loci C11 and SCYOR267c [25]. These two loci were chosen because RC212 is heterozygous and D254 is homozygous at both of these loci, resulting in two fragments for RC212 and one fragment for D254 [14]. Additionally, the size of the two loci was separated by 78 base pairs, which allowed for simultaneous analysis.

2.5. Chemical Analysis

A total of 11 fermentation-derived compounds were selected based on reports of their importance to Pinot Noir, their importance to flavor and aroma, and whether they were yeast strain-dependent. Four of these compounds (ethyl butyrate, isoamyl acetate, 1-hexanol, and phenethyl alcohol) were quantified with gas chromatography mass spectrometry (GC-MS) at UBC Okanagan. A Varian/Agilent CP-3800 GC equipped with a VF-5MS 30 m × 0.25 mm FactorFour capillary column and with a CP-8400 auto sampler was used for a splitless analysis. The injector was ramped from 40 to 100 °C at 10 °C/min. The oven was ramped from 40 to 240 °C at 10 °C/min and a solvent delay of 2.5 min was used. Samples were extracted with liquid-liquid extraction using a 1:1 ratio of the solvents pentane and diethyl ether. A combination of 5 mL sample, 5 µL of 1.615 mg/L methyl isobutyl carbinol (MIBC), and 5 mL solvent were shaken vigorously in large test tubes. The solution settled for 1 h and the extract was transferred from the top layer to GC–MS vials. The other seven compounds (ethyl acetate, acetaldehyde, methanol, 1-propanol, isobutanol, amyl alcohol, and isoamyl alcohol) were quantified by ETS laboratories (St. Helena, CA, USA), using a gas chromatography flame ionization detector (GC-FID), as per the methods of the American Association for Laboratory Accreditation.

2.6. Data Analysis

Strain ratios at the start of the fermentation were compared with expected ratios and with pooled data from subsequent stages by performing a Chi-square goodness of fit test. Relative abundance of strains was compared between treatments and controls by performing a one-way analysis of variance (ANOVA) on data that had fermentation stages pooled, as well as a one-way ANOVA on the end-stage of each treatment. Furthermore, the relative abundance of RC212 in the co-inoculated fermentations was compared between fermentation stages of the same treatment by performing one-way ANOVAs. When significance was indicated, a Tukey–Kramer honest significant difference (HSD) *post-hoc* test was performed. The relationship between the abundance of RC212 and the concentrations of fermentation-derived compounds was determined using regression analysis. Hierarchical cluster analysis, based on Ward's method with euclidean distance, was used to group treatments [9,26]; these results were visualized using a Principal Component Analysis (PCA). The statistical analyses mentioned above were conducted using JMP® 11.0.1. The hierarchical cluster analysis and PCA employed an R 2.0 platform add-in. The concentrations of fermentation-derived chemical compounds were compared between inoculation treatments by performing one-way ANOVAs. When significant differences were detected, Tukey–Kramer HSD *post-hoc* tests were performed to determine differences between treatments. Statistical analysis of chemical compounds was performed using the Rcmdr package in RStudio version 3.1.1. All results were considered significant at $p < 0.05$.

3. Results

The starting proportions of RC212 to D254, sampled immediately after co-inoculation, were not different from their expected ratios (1:1 treatment: $X^2 = 0.563$, $p = 0.453$; 4:1 treatment: $X^2 = 0.039$, $p = 0.844$) (Table 1). This indicated that our inoculation treatments were accurate, which was important in order to make conclusions about the competition between these two strains and about the specificity of chemical compounds to one strain or the other. The co-fermentation treatments differed significantly in their proportion of yeast strains from both control treatments and from each other, when all fermentation stages were pooled (F = 436.1, $p < 0.0001$). Furthermore, the yeast ratios at the end stage of fermentation differed significantly between the two co-fermentation treatments (F = 171.4, $p < 0.0001$), but the yeast proportions of each co-inoculation treatment were constant throughout fermentation (1:1 treatment: F = 0.50, $p = 0.70$; 4:1 treatment: F = 1.6, $p = 0.27$). These results confirm that the proportions of RC212 and D254 differed between all co-inoculation and control treatments at both the beginning and throughout fermentation, and that the inoculated yeast ratios remained constant over the course of fermentation for both co-inoculated treatments.

Table 1. Percent relative abundance of RC212. Chi-square tests were performed to compare pooled data from the early, mid, and end stage ratios with the start ratio of a given treatment. Statistics were only run on the two co-inoculated treatments and not on the pure culture treatments. Any bolded results indicate significance at $p < 0.05$.

RC212:D254		Fermentation Stage			Chi-Square Results	
Treatments	Start	Early	Mid	End	χ^2	*p*-Value
1:1	46 ± 0.07	43 ± 0.11	35 ± 0.07	50 ± 0.09	3.131	0.077
4:1	77 ± 0.02	88 ± 0.04	73 ± 0.04	76 ± 0.04	0.620	0.431
1:0	100 ± 0	100 ± 0	100 ± 0	100 ± 0		
0:1	0 ± 0	0 ± 0	0 ± 0	0 ± 0		

There was a positive linear relationship between the abundance of RC212 and the quantity of four compounds present during fermentations. These compounds were acetaldehyde, 1-propanol, isobutanol, and isoamyl alcohol (Table 2). Alternatively, there was a negative linear relationship between the abundance of RC212 and the quantity of ethyl acetate, amyl alcohol, and isoamyl acetate. We considered that a positive relationship indicated specificity towards RC212 and a negative relationship indicated specificity towards D254. The compounds, ethyl butyrate and phenethyl alcohol, while detected, were not significantly correlated with the relative abundance of RC212 (Table 2). The compounds 1-hexanol and methanol were not detected in any treatment (Table 3).

Table 2. Regression analysis between chemical concentrations and abundances of RC212. Chemicals having a positive linear relationship with RC212 abundance indicate RC212 strain specificity. Chemicals having a negative linear relationship with RC212 abundance indicate D254 specificity. Any bolded results indicate significance at $p < 0.05$.

Chemical	Correlation with RC212	*p*-Value	R^2 Value
Acetaldehyde	+	0.0003	0.740
1-Propanol	+	<0.0001	0.878
Isobutanol	+	<0.0001	0.938
Isoamyl alcohol	+	<0.0001	0.791
Ethyl acetate	−	0.0166	0.452
Amyl alcohol	−	0.0011	0.674
Isoamyl acetate	−	0.0132	0.475
Ethyl butyrate	None	0.2706	0.120
Phenethyl alcohol	None	0.7611	0.010

In our study, RC212 produced significantly higher levels of isobutanol than did D254. Production of this compound by RC212 was also evident in the two co-inoculated treatments, which both contained higher levels of this compound than the pure D254 treatment, but lower levels than the pure RC212 treatment (Table 3). In the pure RC212 cultures, the concentration of this compound approached the sensory threshold of 300 mg/L [18]. For all treatments, isoamyl alcohol was detected at concentrations approaching its bitter sensory threshold of 300 mg/L, although its production was significantly higher in the pure RC212 treatment than the pure D254 treatment (Table 3). Acetaldehyde and 1-propanol concentrations were well below their aroma thresholds of 100–125 mg/L for all treatments [27,28] (Table 3). Ethyl acetate was detected in all treatments at levels above its detection threshold but well below its solvent-like threshold of 100 mg/L, and above the sensory threshold for fruitiness [23]. Unlike previous studies [10,29,30], no strain specificity in the production of ethyl butyrate or phenethyl alcohol was detected (Table 3).

Table 3. Summary of fermentation-derived compounds in concentration (mg/L) for all controlled fermentation treatments. Values are means ± S.E. (*n* = 3). Different superscript letters indicate significant differences between treatments at *p* < 0.05. Each compound was analyzed separately.

Compounds	Treatments (RC212:D254)			
	1:0	4:1	1:1	0:1
Acetaldehyde	22.0 ± 4.6 [a]	17.7 ± 2.5 [a,b]	13.7 ± 1.5 [b,c]	10.0 ± 1.0 [c]
Higher alcohols				
1-Hexanol	ND	ND	ND	ND
Isoamyl alcohol	280.7 ± 6.2 [a]	254.7 ± 10.5 [a,b]	261.3 ± 1.8 [a,b]	249.7 ± 1.8 [b]
Isobutanol	270.7 ± 8.3 [a]	173.0 ± 14.2 [b]	138.3 ± 3.2 [c]	89.0 ± 1.7 [d]
Methanol	ND	ND	ND	ND
Amyl alcohol	47.0 ± 3.5 [a,b]	45.3 ± 4.5 [b]	51.3 ± 1.5 [a,b]	53.7 ± 0.6 [a]
Phenethyl alcohol	23.3 ± 5.5 [a]	27.3 ± 6.1 [a]	18.4 ± 4.1 [a]	22.1 ± 3.5 [a]
1-Propanol	47.3 ± 3.2 [a]	41.0 ± 2.0 [b]	39.0 ± 1.0 [b,c]	34.0 ± 1.0 [c]
Esters				
Ethyl acetate	17.0 ± 3.6 [b]	22.3 ± 5.1 [a,b]	27.0 ± 2.7 [a]	25.3 ± 1.2 [a,b]
Ethyl butyrate	0.49 ± 0.006 [b]	0.53 ± 0.006 [a]	0.49 ± 0.005 [b]	0.50 ± 0.007 [b]
Isoamyl acetate	1.4 ± 0.0 [b]	1.6 ± 0.0 [a]	1.5 ± 0.0 [a,b]	1.5 ± 0.1 [a,b]

ND: not detected.

A Principal Components Analysis (PCA) showed the chemical profile of the 4:1 RC212:D254 co-inoculation treatment clustering with the chemical profile of the pure RC212 fermentations, while the profiles of the 1:1 ratio co-inoculation treatment clustered with the D254 pure culture (Figure 1). The pure RC212 culture fermentations, as well as the 4:1 RC212:D254 co-inoculated fermentations, were correlated with the presence of 1-propanol, acetaldehyde, isobutanol, and isoamyl alcohol. The D254 pure culture fermentations, as well as the 1:1 RC212:D254 co-inoculated fermentations, were correlated with the presence of isoamyl acetate, amyl alcohol, and ethyl acetate.

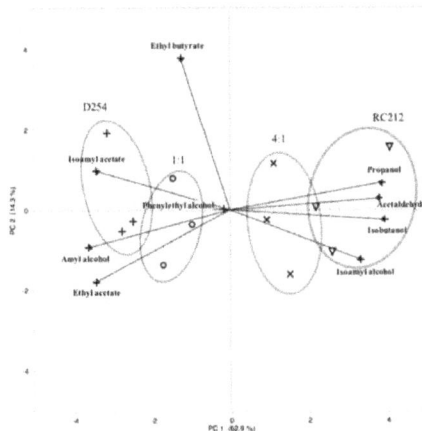

Figure 1. Principal Component Analysis of fermentation-derived compounds detected in each fermentation treatment. The variation (62.9%) among chemical profiles for all treatments can be attributed to a primary principal component (PC1) that differentiates the treatments into two unique chemical groups: (1) D254 pure culture and 1:1 ratio fermentations; and (2) RC212 pure culture and 4:1 ratio fermentations.

4. Discussion

The finding that the proportion of RC212 to D254 remained constant throughout fermentation in both co-inoculation treatments suggests that there was a lack of competitive exclusion under controlled conditions between RC212 and D254, which does not support our first hypothesis that D254 would out-compete RC212 even when RC212 was inoculated in a 4:1 RC212:D254 ratio. Our original hypothesis was based on winery-based studies [3,14], where physical, chemical, and microbial conditions likely differ from in-lab fermentations. We are not aware of any other in-lab studies that have followed the interaction of these two strains during co-fermentation. Nevertheless, one study has followed mixtures of different *S. cerevisiae* strains throughout fermentation and showed both strain exclusion as well as situations where inoculated ratios remained the same throughout fermentation [12]. A second co-inoculation study, using three different commercial strains, observed one strain (Anchor® Vin7) competively excluding Anchor® Vin13 and Lalvin® QA23 [13].

Production of isobutanol was highest in the pure RC212 treatment and lowest in the pure D254 treatment. Thus, the presence of D254 in the co-inoculated treatments appeared to have an inhibiting effect on the production of isobutanol by RC212, as evidenced by the decrease in isobutanol concentration with increasing relative abundance of D254 in the co-fermentations. Although we are not aware of any study that has worked with these two strains in grape must, one other study has shown levels of both n-butanol and isobutanol to differ between some *S. cerevisiae* pure cultures and their mixtures, indicating a significant production trend due to strain interactions [10]. In our RC212 pure cultures, the concentration of isobutanol approached the sensory threshold of 300 mg/L [18], where it could produce bitter flavors; however, solvent-like aromas and flavors probably would not be produced until it neared concentrations of 400 mg/L [18,24]. Although isoamyl alcohol showed the same trend as isobutanol, with respect to pure cultures, the co-inoculations did not result in a significant trend. We are not aware of any studies that have observed the effects of *S. cerevisiae* strain interactions on isoamyl alcohol. As with isobutanol, isoamyl alcohol could produce bitter flavors at the concentrations we found, but not solvent-like aromas and flavors. Both isobutanol and isoamyl alcohol are derivatives of amino acids, so the high concentrations of these compounds were likely, in part, a reflection of the amino acid content in the initial must [9,17]. We did not find a significant interaction trend for isoamyl acetate in both co-inoculation treatments. Our results were similar to another study where ethyl ester concentrations of strain mixtures were similar or slightly higher than those of pure cultures [10]. Supporting our results, this previously conducted study found ethyl esters, including ethyl acetate, above their sensory thresholds for fruity aromas, but not for solvent-like characteristics [10]. Acetaldehyde and 1-propanol concentrations were well below their aroma thresholds of 100–125 mg/L for all treatments, and thus they did not likely contribute directly to the sensorial characteristics of these wines. Many of the compounds we evaluated were below their detection limits, but it is important to note that our study reports on only a small portion of chemicals that are important in contributing to the sensory profile of wine. A full metabolomics study may reveal other chemicals that are important in the interaction of these two strains.

The results of PCA cluster analysis revealed the RC212 pure culture and the 4:1 co-inoculation treatment shared similar chemical profiles, separate from the D254 pure culture and the 1:1 co-inoculation treatment, which also shared similar chemical profiles. This suggests that when the two strains were equally abundant, D254 had a greater effect on the chemical profile than did RC212. This also suggests that the presence of D254 reduced the chemical profile that was contributed by RC212, as evidenced by the reduction in production of a number of chemicals positively correlated with RC212, including acetaldehyde, isobutanol, and 1-propanol. These results support our second hypothesis that D254 would have a greater impact on the chemical profile than RC212 when the cell numbers of the two strains were equal. Nevertheless, our results did not support our third hypothesis that D254 would have a greater impact than RC212 on key fermentation-derived chemicals when the inoculation was administered in a 4:1 RC212:D254 ratio. In our study, both co-inoculated abundance ratios remained constant throughout the fermentations, and when a 4:1 RC212:D254 ratio was in place, the chemical profile resembled the RC212 pure culture more than the D254 pure culture. Nevertheless, the chemical profiles of the co-inoculations shared some of the characteristics of both pure culture fermentations, which supports the results of other studies showing that chemical profiles differ between co-inoculation and pure culture fermentations [9–12]. This indicates that the interaction between two or more strains creates a new chemical profile not seen in the pure cultures. The interactions of multiple strains during fermentation can have synergistic or antagonistic effects on the final sensory attributes of wine [9,12,31], which makes strain selection an important consideration for commercial winemakers. Our results, along with those of Saberi *et al.* [10], suggest that by increasing the number of different strains in a fermentation, a more complex wine, in terms of chemical profile, can be achieved and managed due to multiple interactions between different strains of yeasts. Further research is necessary to determine whether increasing the number of strains in fermentation has an additive effect on the complexity of the wine's chemical profile.

5. Conclusions

In contrast to our original prediction, RC212 and D254 maintained their original inoculation ratios throughout the bench-top fermentations, suggesting that neither RC212 nor D254 competitively excluded the other strain under controlled conditions. The chemical profiles of both co-inoculated fermentations shared some characteristics of each pure culture fermentation. Nevertheless, when the two strains were equally abundant, D254 had a greater impact on the chemical profile than did RC212; this is in support of our hypothesis that D254 would have a relatively greater impact than RC212 on the chemical profile of wine. This is the first report to show that the co-fermentations of these two commercial strains can result in chemical profiles that are different than what is found when each strain is fermenting in pure culture.

Acknowledgments: This work was supported by Quails' Gate Estate Winery and the Natural Sciences and Research Engineering Council (NSERC) through an NSERC Collaborative Research Development (CRD) grant CRDPJ 406796-10, as well as by the UBC internal wine grant. We thank Grant Stanley and David Ledderhof for providing valuable assistance throughout the study. We also graciously thank Lallemand Inc. for the donation of Lalvin® yeasts.

Author Contributions: Frida S. Gustafsson, Vladimir Jiranek, and Daniel M. Durall conceived and designed the experiment. Frida S. Gustafsson conducted the experiment. Marissa Neuner, Chrystal M. Scholl, Sydney C. Morgan, and Daniel M. Durall contributed to the analysis, interpretation, and visualization of data, and the writing of this manuscript.

Conflicts of Interest: The authors declare no conflict of interest.

Abbreviations

The following abbreviations are used in this manuscript:

RC21   Lalvin® Bourgorouge RC212
D254   Lalvin® ICV-D254

References

1. Bisson, L.F.; Joseph, C.M.; Yeasts, L. *Biology of Microorganisms on Grapes, in Must and in Wine*; Konig, H., Unden, G., Frohlich, J., Eds.; Springer Berlin Heidelberg: Berlin, Germany, 2009; pp. 47–60.
2. Clavijo, A.; Calderón, I.L.; Paneque, P. Effect of the Use of Commercial *Saccharomyces* Strains in a Newly Established Winery in Ronda (Málaga, Spain). *Int. J. Gen. Mol. Microbiol.* **2011**, *99*, 727–731. [CrossRef] [PubMed]
3. Lange, J.N.; Faasse, E.; Tantikachornkiat, M.; Gustafsson, F.S.; Halvorsen, L.C.; Kluftinger, A.; Ledderhof, D.; Durall, D.M. Implantation and Persistence of Yeast Inoculum in Pinot Noir Fermentations at Three Canadian Wineries. *Int. J. Food Microbiol.* **2014**, *180*, 56–61. [CrossRef] [PubMed]
4. Lee, P.-R.; Saputra, A.; Yu, B.; Curran, P.; Liu, S.-Q. Effects of Pure and Mixed-Cultures of *Saccharomyces cerevisiae* and *Williopsis saturnus* on the Volatile Profiles of Grape Wine. *Food Biotechnol.* **2012**, *26*, 307–325. [CrossRef]

5. Sadoudi, M.; Tourdot-Maréchal, R.; Rousseaux, S.; Steyer, D.; Gallardo-Chacón, J.J.; Ballester, J.; Vichi, S.; Guérin-Schneider, R.; Caixach, J.; Alexandre, H. Yeast-Yeast Interactions Revealed by Aromatic Profile Analysis of Sauvignon Blanc Wine Fermented by Single or Co-Culture of Non-*Saccharomyces* and *Saccharomyces* Yeasts. *Food Microbiol.* **2012**, *32*, 243–253. [CrossRef] [PubMed]

6. Comitini, F.; Gobbi, M.; Domizio, P.; Romani, C.; Lencioni, L.; Mannazzu, I.; Ciani, M. Selected Non-*Saccharomyces* Wine Yeasts in Controlled Multistarter Fermentations with *Saccharomyces cerevisiae*. *Food Microbiol.* **2011**, *28*, 873–882. [CrossRef] [PubMed]

7. Domizio, P.; Romani, C.; Lencioni, L.; Comitini, F.; Gobbi, M.; Mannazzu, I.; Ciani, M. Outlining a Future for Non-*Saccharomyces* Yeasts: Selection of Putative Spoilage Wine Strains to Be Used in Association with *Saccharomyces cerevisiae* for Grape Juice Fermentation. *Int. J. Food Microbiol.* **2011**, *147*, 170–180. [CrossRef] [PubMed]

8. Viana, F.; Gil, J.V.; Genovés, S.; Vallés, S.; Manzanares, P. Rational Selection of Non-*Saccharomyces* Wine Yeasts for Mixed Starters Based on Ester Formation and Enological Traits. *Food Microbiol.* **2008**, *25*, 778–785. [CrossRef] [PubMed]

9. King, E.S.; Kievit, R.L.; Curtin, C.; Swiegers, J.H.; Pretorius, I.S.; Bastian, S.E.P.; Francis, I.L. The Effect of Multiple Yeasts Co-Inoculations on Sauvignon Blanc Wine Aroma Composition, Sensory Properties and Consumer Preference. *Food Chem.* **2010**, *122*, 618–626. [CrossRef]

10. Saberi, S.; Cliff, M.A.; van Vuuren, H.J.J. Impact of Mixed *S. cerevisiae* Strains on the Production of Volatiles and Estimated Sensory Profiles of Chardonnay Wines. *Food Res. Int.* **2012**, *48*, 725–735. [CrossRef]

11. Barrajón, N.; Arévalo-Villena, M.; Úbeda, J.; Briones, A. Enological Properties in Wild and Commercial *Saccharomyces cerevisiae* Yeasts: Relationship with Competition during Alcoholic Fermentation. *World J. Microbiol. Biotechnol.* **2011**, *27*, 2703–2710. [CrossRef]

12. Howell, K.S.; Cozzolino, D.; Bartowsky, E.J.; Fleet, G.H.; Henschke, P.A. Metabolic Profiling as a Tool for Revealing *Saccharomyces* Interactions during Wine Fermentation. *FEMS Yeast Res.* **2006**, *6*, 91–101. [CrossRef] [PubMed]

13. King, E.S.; Swiegers, J.H.; Travis, B.; Francis, I.L.; Bastian, S.E.P.; Pretorius, I.S. Coinoculated Fermentations Using *Saccharomyces* Yeasts Affect the Volatile Composition and Sensory Properties of *Vitis vinifera* L. cv. Sauvignon Blanc Wines. *J. Agric. Food Chem.* **2008**, *56*, 10829–10837. [CrossRef] [PubMed]

14. Hall, B.; Durall, D.M.; Stanley, G. Population Dynamics of *Saccharomyces cerevisiae* during Spontaneous Fermentation at a British Columbia Winery. *Am. J. Enol. Vitic.* **2011**, *62*, 66–72. [CrossRef]

15. Lange, J.N. Yeast Population Dynamics During Inoculated and Spontaneous Fermentations at Three Local British Columbia Wineries. Master Thesis, University of British Columbia, Kelowna, Canada, December 2012.

16. Bisson, L.F.; Karpel, J.E. Genetics of Yeast Impacting Wine Quality. *Annu. Rev. Food Sci. Technol.* **2010**, *1*, 139–162. [CrossRef] [PubMed]

17. Jiranek, V.; Langridge, P.; Henschke, P.A. Amino Acid and Ammonium Utilization by *Saccharomyces cerevisiae* Wine Yeasts from a Chemically Defined Medium. *Am. J. Enol. Vitic.* **1995**, *46*, 75–83.

18. Swiegers, J.H.; Bartowsky, E.J.; Henschke, P.A.; Pretorius, I.S. Yeast and Bacterial Modulation of Wine Aroma and Flavour. *Aust. J. Grape Wine Res.* **2005**, *11*, 139–173. [CrossRef]

19. Romano, P.; Fiore, C.; Paraggio, M.; Caruso, M.; Capece, A. Function of Yeast Species and Strains in Wine Flavour. *Int. J. Food Microbiol.* **2003**, *86*, 169–180. [CrossRef]

20. Styger, G.; Prior, B.; Bauer, F.F. Wine Flavor and Aroma. *J. Ind. Microbiol. Biotechnol.* **2011**, *38*, 1145–1159. [CrossRef] [PubMed]

21. Suárez-Lepe, J.A.; Morata, A. New Trends in Yeast Selection for Winemaking. *Trends Food Sci. Technol.* **2012**, *23*, 39–50. [CrossRef]

22. Ugliano, M.; Henschke, P.A. Yeasts and Wine Flavour. In *Wine Chemistry and Biochemistry*; Moreno-Arribas, M.V., Polo, M.C., Eds.; Springer: New York, NY, USA, 2009; pp. 313–374.

23. Sumby, K.M.; Grbin, P.R.; Jiranek, V. Microbial Modulation of Aromatic Esters in Wine: Current Knowledge and Future Prospects. *Food Chem.* **2010**, *121*, 1–16. [CrossRef]

24. Ribereau-Gayon, P.; Maujean, A.; Dubourdieu, D. *Handbook of Enology*, 2nd ed.; John Wiley and Sons Ltd.: Chichester, UK, 2006.

25. Legras, J.L.; Ruh, O.; Merdinoglu, D.; Karst, F. Selection of Hypervariable Microsatellite Loci for the Characterization of *Saccharomyces cerevisiae* Strains. *Int. J. Food Microbiol.* **2005**, *102*, 73–83. [CrossRef] [PubMed]

26.   Capece, A.; Romaniello, R.; Poeta, C.; Siesto, G.; Massari, C.; Pietrafesa, R.; Romano, P. Control of Inoculated Fermentations in Wine Cellars by Mitochondrial DNA Analysis of Starter Yeast. *Ann. Microbiol.* **2011**, *61*, 49–56. [CrossRef]

27.   Grosch, W. Evaluation of the Key Odorants of Foods by Dilution Experiments, Aroma Models and Omission. *Chem. Senses* **2001**, *26*, 533–545. [CrossRef] [PubMed]

28.   Liu, S.-Q.; Pilone, G.J. An Overview of Formation and Roles of Acetaldehyde in Winemaking with Emphasis on Microbiological Implications. *Int. J. Food Sci. Technol.* **2000**, *35*, 49–61. [CrossRef]

29.   Lilly, M.; Bauer, F.F.; Styger, G.; Lambrechts, M.G.; Pretorius, I.S. The Effect of Increased Branched-Chain Amino Acid Transaminase Activity in Yeast on the Production of Higher Alcohols and on the Flavour Profiles of Wine and Distillates. *FEMS Yeast Res.* **2006**, *6*, 726–743. [CrossRef] [PubMed]

30.   Saerens, S.M.G.; Delvaux, F.; Verstrepen, K.J.; Van Dijck, P.; Thevelein, J.M.; Delvaux, F.R. Parameters Affecting Ethyl Ester Production by *Saccharomyces cerevisiae* during Fermentation. *Appl. Environ. Microbiol.* **2008**, *74*, 454–461. [CrossRef] [PubMed]

31.   Favale, S.; Pietromarchi, P.; Ciolfi, G. Metabolic Activity and Interactions between Two Strains, *Saccharomyces cerevisiae* r.f. *bayanus* (SBC2) and *Saccharomyces cerevisiae* r.f. *uvarum* (S6u), in Pure and Mixed Culture Fermentations. *Vitis J. Grapevine Res.* **2007**, *46*, 39–43.

fermentation

MDPI

Article

A Bacterial Laccase for Enhancing Saccharification and Ethanol Fermentation of Steam-Pretreated Biomass

Antonio D. Moreno [1,2], David Ibarra [3,*], Antoine Mialon [4] and Mercedes Ballesteros [2]

[1]   IMDEA Energía, Biotechnological Processes for Energy Production Unit, Móstoles, Madrid 28935, Spain; david.moreno@ciemat.es

[2]   CIEMAT, Renewable Energy Division, Biofuels Unit, Avda. Complutense 40, Madrid 28040, Spain; m.ballesteros@ciemat.es

[3]   INIA-CIFOR, Forestry Products Department, Cellulose and Paper Laboratories, Ctra de La Coruña Km 7.5, Madrid 28040, Spain

[4]   MetGen Oy, Rakentajantie 26, Kaarina 20780, Finland; antoine@metgen.com

*   Correspondence: ibarra.david@inia.es; Tel.: +34-91-347-3948

Academic Editor: Ronnie G. Willaert

Received: 8 April 2016; Accepted: 26 April 2016; Published: 4 May 2016

Abstract: Different biological approaches, highlighting the use of laccases, have been developed as environmentally friendly alternatives for improving the saccharification and fermentation stages of steam-pretreated lignocellulosic biomass. This work evaluates the use of a novel bacterial laccase (MetZyme) for enhancing the hydrolysability and fermentability of steam-exploded wheat straw. When the water insoluble solids (WIS) fraction was treated with laccase or alkali alone, a modest increase of about 5% in the sugar recovery yield (glucose and xylose) was observed in both treatments. Interestingly, the combination of alkali extraction and laccase treatment boosted enzymatic hydrolysis, increasing the glucose and xylose concentration in the hydrolysate by 21% and 30%, respectively. With regards to the fermentation stage, the whole pretreated slurry was subjected to laccase treatment, lowering the phenol content by up to 21%. This reduction allowed us to improve the fermentation performance of the thermotolerant yeast *Kluyveromyces marxianus* CECT 10875 during a simultaneous saccharification and fermentation (SSF) process. Hence, a shorter adaptation period and an increase in the cell viability—measured in terms of colony forming units (CFU/mL)—could be observed in laccase-treated slurries. These differences were even more evident when a presaccharification step was performed prior to SSF. Novel biocatalysts such as the bacterial laccase presented in this work could play a key role in the implementation of a cost-effective technology in future biorefineries.

Keywords: alkaline extraction; bacterial Metzyme laccase; lignocellulosic ethanol; simultaneous saccharification and fermentation; thermotolerant yeast

1. Introduction

The transition towards a post-petroleum society for mitigating global climate change is currently led by the development and implementation of biorefineries. Biorefineries will be competitive, innovative and sustainable local industries for the production of plant- and waste-derived fuels, materials and chemicals. Due to its low costs and wide distribution, lignocellulosic biomass is the most promising feedstock to be used in biorefineries, and lignocellulose-derived fuels, including ethanol, the most significant product.

Many different feedstocks, conversion methods, and process configurations have been studied for lignocellulosic ethanol production, with the biochemical route being the most promising option [1]. Lignocellulose is a complex matrix where a 'skeleton' polymer, cellulose, is coated by two 'protective' polymers, hemicellulose and lignin. Biochemical conversion of lignocellulosic biomass includes a pretreatment step to open up the structure and increase biomass digestibility. Subsequently, cellulose and hemicellulose polymers are subjected to an enzymatic saccharification process to obtain the fermentable sugars. The optimal performance of cellulolytic enzymes is therefore a crucial step that determines the overall process efficiency. Finally, the resulting sugars are converted into ethanol via microbial fermentation [1].

Pretreatment influences lignocellulose digestibility by an extensive modification of the structure. A large number of pretreatment technologies, mainly physical and/or chemical, have been developed and applied on a wide variety of feedstocks [2]. Among them, hydrothermal pretreatments, such as steam explosion, are considered the most effective methods and are commonly used for lignocellulose-to-ethanol conversion. The action mechanism of these pretreatment technologies lies in the solubilisation of hemicellulose fraction and the redistribution and/or modification of lignin, which increase outstandingly the hydrolysis of cellulose without the need of adding any catalyst [3]. These pretreatment technologies, however, still present several drawbacks that must be overcome. First, the residual lignin that is left in the pretreated materials represents an important limiting factor during the enzymatic hydrolysis of carbohydrates, promoting the non-specific adsorption of hydrolytic enzymes and, in turn, decreasing saccharification yields [4]. Second, these pretreatment methods generate some soluble compounds, derived from sugar degradation (furan derivatives and weak acids) and partial lignin solubilisation (aromatic acids, alcohols and aldehydes), which inhibit cellulolytic enzymes and fermentative microorganisms [5]. Performing a delignification step prior to the addition of hydrolytic enzymes may reduce the non-productive adsorption of these enzymes, enhancing the saccharification yields. In the same way, a detoxification process may reduce the amount of inhibitors produced after steam explosion pretreatment, boosting the saccharification and fermentation steps. Different physico/chemical technologies have been studied for delignification and detoxification of pretreated materials [2,6]. However, most of these methods require extra equipment and additional steps and have high energy demands, complicating the lignocellulose-to-ethanol process and increasing the production costs. As an alternative to physico/chemical methods, the use of ligninolytic enzymes such as laccases may provide further integration into the process and lower energy requirements [7].

Laccases are multicopper oxidases that catalyze the one-electron oxidation of phenols, anilines and aromatic thiols to their corresponding radicals with the concomitant reduction of molecular oxygen to water. Laccases are mainly produced by plants and fungi, including the white-rot basidiomycetes responsible for lignin degradation in nature [8]. Also, some bacterial laccases have been described and fully characterized, generally showing lower redox potential and more stable at high pH and temperatures compared to fungal laccases [9]. The role of laccases in lignin degradation makes them attractive biocatalysts for the pulp and paper industry as substitutes of chlorine-containing reagents in pulp bleaching [10,11]. Both fungal and bacterial laccases have been studied with beneficial results [12,13]. Moreover, they are used in wastewater treatment to detoxify industrial effluents with high phenolic content—such as the streams obtained during pulp and paper production—due to their ability to oxidize phenolic compounds [14,15].

The vast experience gained from the extensive use of laccases in the paper pulp industry has provided an excellent starting point for the application of laccases within a broader perspective. In this context, different fungal laccases have been widely studied for improving the conversion efficiency of lignocellulose into ethanol, and consequently increasing final product concentrations [7,16–25]. Nevertheless, little is known about the use of bacterial laccases for these purposes. The present work evaluates the commercial bacterial laccase MetZyme, exploring its potential for improving the hydrolysability and fermentability of steam-exploded wheat straw.

2. Materials and Methods

2.1. Raw Material and Steam Explosion Pretreatment

Wheat straw, supplied by CEDER-CIEMAT (Soria, Spain), was used as raw material. It presented the following composition (% dry weight (DW)): cellulose, 40.5 ± 2.1; hemicelluloses, 26.1 ± 1.1 (xylan, 22.7 ± 0.5; arabinan 2.1 ± 0.4; and galactan, 1.3 ± 0.2); Klason lignin, 18.1 ± 0.8; ashes, 5.1 ± 0.3; and extractives, 14.6 ± 0.4.

Prior to steam explosion, wheat straw was milled in a laboratory hammer mill to obtain a chip size between 2 and 10 mm. Then, the milled material was pretreated in a 10 L reactor at 200 °C for 2.5 min. The recovered slurry was handled differently depending on its further use. For analytical purposes, one portion was vacuum filtered with the aim of obtaining a liquid fraction or prehydrolysate and a solid fraction. Subsequently, the solid fraction was thoroughly washed with distilled water to obtain the water insoluble solids (WIS) fraction. Chemical composition of both raw and pretreated material (WIS fraction) was determined using the Laboratory Analytical Procedures (LAP) for biomass analysis, provided by the National Renewable Energies Laboratory [26]. Sugars and degradation compounds contained in the liquid fraction were also measured. Most of the sugars present in the liquid fraction were in oligomeric form, and therefore a mild acid hydrolysis (4% (*v/v*) H_2SO_4, 120 °C and 30 min) was required to determine the concentration of monomeric sugars. The obtained WIS fraction was also used for saccharification studies since the majority of the inhibitory compounds were removed. On the other hand, the remained slurry was used as substrate to evaluate its fermentability due to the higher inhibitor content. Both WIS and slurry were stored at 4 °C until their use.

2.2. Enzymes

An industrial thermostable bacterial laccase (pH range 3–8) was specifically selected from MetGen's products portfolio (MetZyme, Cat.-No: 10-101-UF, MetGen Oy, Kaarina, Finland), and used in both saccharification and fermentation assays. Laccase activity (284 IU/g of laccase activity) was measured by oxidation of 5 mM 2,2′-azino-bis(3-ethylbenzothiazoline-6-sulphonic acid) (ABTS) to its cation radical in 0.1 M sodium acetate (pH 5) at 24 °C. Formation of the ABTS cation radical was monitored at 436 nm ($\varepsilon_{436} = 29{,}300\ \text{M}^{-1} \cdot \text{cm}^{-1}$).

A mixture of NS50013 and NS50010, both produced by Novozymes (Bagsvaerd, Denmark), was used for the saccharification of steam-pretreated what straw. NS50013 (60 FPU/mL of cellulase activity) is a cellulase preparation that presents low β-glucosidase activity and therefore it requires the supplementation with NS50010 (810 IU/mL of β-glucosidase activity), which mainly presents β-glucosidase activity. Overall cellulase activity was determined using filter paper (Whatman No. 1 filter paper strips), while β-glucosidase activity was measured using cellobiose as a substrate. The enzymatic activities were followed by the release of reducing sugars [27].

One unit of enzyme activity was defined as the amount of enzyme that transforms 1 μmol of substrate per minute.

2.3. Microorganism and Growth Conditions

Kluyveromyces marxianus CECT 10875, a thermotolerant strain selected by Ballesteros *et al.* [28], was employed as fermentative microorganism in this study. Active cultures for inoculation were obtained in 100-mL flasks with 50 mL of growth medium containing 30 g/L glucose, 5 g/L yeast extract, 2 g/L NH_4Cl, 1 g/L KH_2PO_4, and 0.3 g/L $MgSO_4 \cdot 7H_2O$. After 16 h on an orbital shaker at 150 rpm and 42 °C, the precultures were centrifuged at 9000 rpm for 10 min. Supernatant was discarded and cells were washed once with distilled water and diluted accordingly to obtain an inoculum level of 1 g/L DW.

2.4. Laccase Treatment and Saccharification of the WIS Fraction

The WIS fraction obtained after steam explosion (200 °C, 2.5 min) was subjected to a sequential laccase treatment and saccharification directly (Strategy 1) or after a mild alkaline extraction (Strategy 2).

Strategy 1, *sequential laccase treatment and saccharification*: 2.5 g DW of the corresponding WIS fraction were suspended in 50 mM sodium citrate buffer (pH 5.5) in 100-mL shake flasks to reach a final concentration of 5% (*w/v*) total solids (TS). This solution was treated with MetZyme laccase (10 IU/g DW substrate) for 24 h at 50 °C and 150 rpm in an orbital shaker. After 24 h of laccase treatment, solids were filtered through a Büchner funnel, washed with 1 L of water and dried at 60 °C. In a subsequent step, the laccase-treated WIS fraction was resuspended with 50 mM sodium citrate buffer (pH 5.5) in 100-mL flasks to reach a final concentration of 5% TS (*w/v*). Solids were subjected to saccharification at 50 °C for 72 h in an orbital shaker (150 rpm), with an enzyme loading of 5 FPU/g DW substrate of NS50013 and 5 IU/g DW substrate of NS50010.

Strategy 2, *mild alkaline extraction and sequential laccase treatment and saccharification*: 2.5 g DW of the corresponding WIS fraction was extracted with alkali (2.5% NaOH, for 1 h at 60 °C and 5% TS (*w/v*) substrate loading) followed by filtration and water washing. Then, the alkali-extracted WIS fraction was resuspended in 50 mM sodium citrate buffer (pH 5.5) in 100-mL flasks to reach a final concentration of 5% TS (*w/v*) and subjected to sequential laccase treatment and saccharification as explained above.

The effects of bacterial laccase treatments on both WIS fractions were evaluated in terms of (1) chemical composition and (2) saccharification yields. The chemical composition of laccase-treated WIS, subjected or not to a mild alkaline extraction, was determined using the NREL-LAP for biomass analysis [26]. On the other hand, the enzymatic hydrolysates obtained from laccase-treated WIS (with and without a previous mild alkaline extraction step) were centrifuged to remove the remaining solids, and the supernatants were analyzed to determine glucose and xylose concentration. For a better comparison between assays, relative glucose/xylose recoveries (RGR; RXR) were calculated as following Equation (1):

$$\text{RGR (\%)} = \text{g/L glucose}_{assay} \times 100/\text{g/L glucose}_{control} \qquad (1)$$

For RXR (%), similar equation was used but with xylose concentration instead.

Control assays were performed under same conditions in Strategy 1 and Strategy 2 without the addition of MetZyme laccase. All the experiments were carried out in triplicate.

2.5. Laccase Treatment and Fermentation of the Whole Slurry

The whole slurry obtained after steam explosion (200 °C, 2.5 min) was subjected to laccase treatment and simultaneous saccharification and fermentation without (Strategy 3) and with (Strategy 4) a presaccharification step to evaluate its fermentability.

Strategy 3, *consecutive laccase treatment and simultaneous saccharification and fermentation (LSSF)*: 2.5 g DW of the corresponding slurry was suspended with 50 mM sodium citrate buffer (pH 5.5) in 100-mL flasks to reach a final concentration of 10% TS (*w/v*). Then, 10 IU/g DW substrate of MetZyme laccase were added and the mixture was incubated at 50 °C and 150 rpm in an orbital shaker for 24 h. After laccase treatment, the slurries were subsequently subjected to a simultaneous saccharification and fermentation (SSF) process at 42 °C for 72 h in an orbital shaker (150 rpm). Laccase-treated slurries were subjected to SSF after the supplementation with 15 FPU/g DW substrate of NS50013, 15 IU/g DW substrate of NS50010, nutrients (those described for cell propagation, except glucose) and 1 g/L DW of *K. marxianus*.

Strategy 4, *consecutive laccase treatment with presaccharification and simultaneous saccharification and fermentation (LPSSF)*: 2.5 g DW of the corresponding slurry were suspended in 50 mM sodium citrate buffer (pH 5.5) in 100-mL flasks to reach a final concentration of 10% TS (*w/v*). Then, 10 IU/g DW substrate of MetZyme laccase were added and the mixture was incubated at 50 °C and 150 rpm in an orbital shaker. After 16 h of laccase treatment, a presaccharification step was carried out for 8 h by supplementing the slurries with 15 FPU/g DW substrate of NS50013 and 15 IU/g DW substrate of NS50010. Afterwards, the temperature was reduced to 42 °C and nutrients and 1 g/L DW of *K. marxianus* were added, which turned the process into a SSF. The experiments were run for another 72 h.

The effect of MetZyme laccase on specific inhibitory compounds was evaluated before yeast addition, *i.e.*, right after laccase treatment or laccase treatment with presaccharification. For that, prior starting SSF processes samples were taken and centrifuged, and the supernatants were analyzed for the identification and quantification of inhibitory compounds. In the same way, samples were periodically withdrawn during SSF processes to determine cell viability and glucose and ethanol concentration (a centrifugation step was included prior to analyze glucose and ethanol concentration).

Control assays were performed under the same conditions without the addition of MetZyme laccase. All the experiments were carried out in triplicate.

2.6. Analytical Methods

Ethanol was analyzed by gas chromatography, using a 7890A GC System (Agilent, Waldbronn, Germany) equipped with an Agilent 7683B series injector, a flame ionization detector and a Carbowax 20 M column operating at 85 °C. Injector and detector temperature was maintained at 175 °C.

Sugar concentration was quantified by high-performance liquid chromatography (HPLC) in a Waters chromatograph equipped with a refractive index detector (Waters, Mildford, MA, USA). A CarboSep CHO-682 carbohydrate analysis column (Transgenomic, San Jose, CA, USA) operating at 80 °C with ultrapure water as a mobile-phase (0.5 mL/min) was used for the separation.

Furfural and 5-hydroxymethylfurfural (5-HMF) were analyzed and quantified by HPLC (Agilent, Waldbronn, Germany), using a Coregel 87H3 column (Transgenomic, San Jose, CA, USA) at 65 °C equipped with a 1050 photodiode-array detector (Agilent, Waldbronn, Germany). As mobile phase, 89% 5 mM H_2SO_4 and 11% acetonitrile at a flow rate of 0.7 mL/min were used.

Formic acid and acetic acid were also quantified by HPLC (Waters, Mildford, MA, USA) using a 2414 refractive index detector (Waters, Mildford, MA, USA) and a Bio-Rad Aminex HPX-87H (Bio-Rad Labs, Hercules, CA, USA) column maintained at 65 °C with a mobile phase (5 mmol/L H_2SO_4) at 0.6 mL/min of flow rate.

Total phenolic content was analyzed according to the Folin-Ciocalteu procedure [29]. 0.5 mL of sample and the serial standard solution (gallic acid) were introduced into test tubes with 2.5 mL of Folin-Ciocalteu's reagent (1:10 dilution in water) and 2 mL of sodium carbonate (7.5% w/v). The tubes were incubated for 5 min at 50 °C. After cooling down the temperature, the absorbance was measured at 760 nm using a Lambda 365 spectrophotometer (PerkinElmer, Boston, MA, USA).

Cell viability was measured by cell counting using agar plates (30 g/L glucose, 5 g/L yeast extract, 2 g/L NH_4Cl, 1 g/L KH_2PO_4, and 0.3 g/L $MgSO_4 \cdot 7H_2O$, 20 g/L agar). Plates were incubated at 42 °C for 24 h.

All analytical values were calculated from duplicates or triplicates. Statistical analyses were performed using IBM SPSS Statistics v22.0 for MacOs X Software (SPSS, Inc., Chicago, IL, USA). The mean and standard deviation were calculated for descriptive statistics. When appropriate, analysis of variance (ANOVA) with or without Bonferroni's post-test was used for comparisons between assays. The level of significance was set at $p < 0.05$, $p < 0.01$ or $p < 0.001$.

3. Results and Discussion

3.1. Pretreated Biomass Composition

Steam explosion pretreatment was performed at 200 °C and 2.5 min (Table 1). In comparison to the cellulose content of the untreated wheat straw (40.5%), steam explosion increased the cellulose proportion of the WIS fraction (53.5%) due to an extensive hemicellulose solubilization and degradation. This solubilization is evidenced by the lower proportion of the remaining hemicellulose (11.7%) fraction of the WIS residue and the high xylose content (32 g/L) in the liquid fraction. Also, different degradation products were recovered in the liquid fraction due to biomass degradation. The most predominant inhibitors were acetic acid, formic acid, furfural, 5-HMF and phenols (Table 1). Acetic acid is formed by the hydrolysis of acetyl groups contained in the hemicellulose structure. Formic acid derives from furfural and 5-HMF degradation, which in turn, results from pentoses (mainly xylose) and hexoses degradation, respectively. Finally, phenols are released during lignin partial solubilization and degradation of lignin [5,30]. A wide variety of phenolic substituted compounds such as 4-hydroxybenzaldehyde, vanillin, syringaldehyde, *p*-coumaric acid or ferulic acid, have been identified in steam-exploded wheat straw [3,31].

Table 1. Composition of steam-exploded wheat straw at 200 °C, 2.5 min.

WIS			Liquid Fraction			
Component	% DW	Sugar Component	Monomeric Form (g/L)	Oligomeric Form (g/L)	Inhibitors	g/L
Cellulose	53.5 ± 1.1	Glucan	2.3 ± 0.2	12.4 ± 0.3	Furfural	0.8 ± 0.0
Hemicellulose	11.7 ± 0.7	Xylan	2.8 ± 0.1	29.2 ± 0.7	5-HMF	0.3 ± 0.0
Lignin	30.4 ± 3.2	Arabinan	1.3 ± 0.2	1.1 ± 0.0	Acetic acid	6.9 ± 0.3
		Galactan	0.4 ± 0.1	1.4 ± 0.1	Formic acid	6.3 ± 0.2
		Mannan	n.d.	n.d.	Phenols	5.9 ± 0.8

5-HMF, 5-hydroxymethylfurfural; n.d., not determined.

3.2. Laccase Treatment and Saccharification of the WIS Fraction

The WIS fraction obtained after pretreatment of wheat straw at 200 °C, 2.5 min was subjected to laccase treatment and saccharification with and without a mild alkaline extraction: Strategy 1, sequential laccase treatment and saccharification; and Strategy 2, mild alkaline extraction and sequential laccase treatment and saccharification.

3.2.1. Effect of Bacterial Laccase Treatment on the Chemical Composition of WIS

The chemical composition of laccase-treated WIS, without and with a previous mild alkaline extraction step, was determined and compared with their respective controls (Table 2). In the case of those pretreated materials that were not subjected to an alkaline extraction, no relevant changes in the lignin content were observed after treatment with MetZyme laccase. Contradictory results have been described with fungal laccases on steam-pretreated materials. Moilanen *et al.* [21] obtained no substantial variation in the lignin content after laccase (*Cerrena unicolor*) treatment of steam-pretreated giant reed (*Arundo donax*). Similar results were obtained by Martín-Sampedro *et al.* [20,32] when steam-exploded eucalypt was treated with *Myceliophtora thermophila* laccase. In contrast, Oliva-Taravilla *et al.* [33] observed a slight increment in the lignin content of unwashed steam-exploded wheat straw after treatment with *Pycnoporus cinnabarinus* laccase. Likewise, Moilanen *et al.* [21] also described a lignin content increment in steam-pretreated spruce (*Picea abis*) treated with *C. unicolor* laccase.

Table 2. Composition of WIS samples treated with bacterial MetZyme laccase without or with a prior alkaline extraction.

Assay	Composition (% DW, w/w) [a]		
	Cellulose	Hemicellulose	Lignin
C	53.5 ± 1.1	11.7 ± 0.7	30.4 ± 3.2
L	54.1 ± 0.6	11.2 ± 0.7	30.5 ± 3.1
Alk + C	58.2 ± 2.9	12.3 ± 0.5	27.6 ± 1.6
Alk + L	57.9 ± 2.7	12.4 ± 0.6	27.1 ± 0.6

[a] The remaining percent (of the whole 100%) for biomass composition is represented by other components, including ashes and acid soluble lignin. C, control without alkaline extraction; ALK + C, control with alkaline extraction; L, laccase treatment, Alk + L, alkaline extraction and laccase treatment. Analysis of variance (ANOVA) with Bonferroni's post-test was performed to identify differences between C, L, Alk + C or Alk + L. Differences in means are not statistically significant.

It is known that alkaline treatment of steam-exploded materials decreases lignin content considerably [23,34,35]. In our study, the alkaline treatment was performed at mild conditions and caused 9% delignification (the mean difference is not significant at the 0.05 level) of steam-exploded wheat straw. When MetZyme laccase treatment was subsequently applied to the alkali-extracted WIS, no benefit were found by combining both treatments and similar values of delignification (11%) were observed (Table 2).

3.2.2. Effect of Bacterial Laccase Treatment on Saccharification Yields

RGRs and RXRs obtained after the saccharification of the WIS fractions treated with laccase are shown in Figure 1. In the case of Strategy 1 (sequential laccase treatment and saccharification), RGR of laccase-treated assays was increased by almost 5% (the mean difference is not significant at the 0.05 level) compared to control hydrolysates (Figure 1A). Similarly, an increment on RXR (3%, the mean difference is not significant at the 0.05 level) was also observed (Figure 1B). Even though no major changes were observed in the lignin content after treatment with this bacterial laccase, the slightly better saccharification yields could be attributed to the modification of the lignin structure on the WIS surface, which would affect the interaction of hydrolytic enzymes with the pretreated material. In this context, the action mechanism of laccases towards phenolic lignin units is altering the hydrophobicity of lignin and, consequently, lowering the non-specific adsorption of cellulases to this polymer. Palonen and Viikari [24] reported an increment of carboxyl groups of lignin from steam-pretreated spruce after treatment with the fungal *T. hirsuta* laccase, decreasing the hydrophobicity of lignin and increasing surface charge. These changes reduced the non-specific adsorption of hydrolytic enzymes on lignin, enhancing saccharification yields. Similar results were also obtained by Moilanen *et al.* [21] when steam-pretreated spruce was treated with *C. unicolor* laccase. Nevertheless, these authors also reported an increase in the non-specific adsorption of cellulases and lower glucose recovery yields when laccase treatment was performed on steam-pretreated giant reed. Oliva-Taravilla *et al.* [33] also described lower saccharification yields when steam-exploded wheat straw was treated with the fungal *P. cinnabarinus* laccase. In that work, the increment in Klason lignin observed in laccase-treated WIS was related to a grafting phenomenon of soluble phenols onto the lignin polymer, which hinders the accessibility of cellulolytic enzymes to cellulose and therefore reduces sugar recoveries.

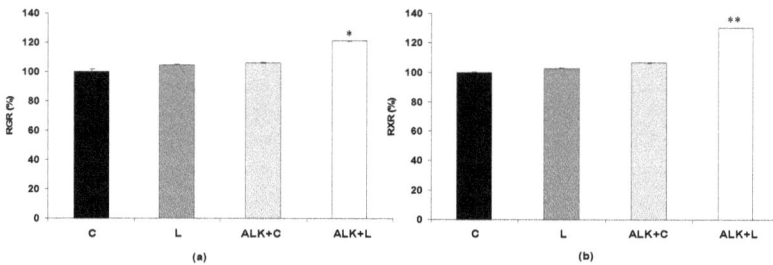

Figure 1. Relative glucose (RGR) (**a**) and xylose (RXR) (**b**) recoveries at 72 h of enzymatic hydrolysis of WIS samples resulting from the different MetZyme laccase treatment and saccharification strategies. Strategy 1, sequential laccase treatment and saccharification (C, control sample; L, laccase sample). Strategy 2, alkaline extraction and sequential laccase treatment and saccharification (ALK + C, control sample with alkaline extraction; ALK + L, laccase sample with alkaline extraction). Glucose concentration values after 72 h of saccharification of control samples were 13.1 and 13.9 g/L for strategies 1 and 2, respectively. Xylose concentration values after 72 h of saccharification of control samples were 2 and 2.2 g/L for strategies 1 and 2, respectively. Mean values and standard deviations were calculated from the triplicates to present the results. Analysis of variance (ANOVA) with Bonferroni's post-test was performed to identify differences between C, L, Alk + C or Alk + L. The mean difference is significant at the (*) 0.05 or (**) 0.01 level.

In the case of Strategy 2 (mild alkaline extraction and sequential laccase treatment and saccharification), the enzymatic hydrolysis of control assays extracted with alkali produced higher RGR (6%, the mean difference is not significant at the 0.05 level) and RXR (7%, the mean difference is not significant at the 0.05 level) values than the control assays not subjected to mild alkaline treatment (Figure 1). This enhancement in saccharification yields after the extraction with alkali is very well known [23,34,35]. Alkali extraction generates new irregular pores as a result of the removal of lignin and the disruption of lignin-carbohydrate complexes, contributing to an increase in the accessibility and susceptibility of cellulose and hemicellulose polymers to the action of hydrolytic enzymes. These advantages can be boosted by a subsequent laccase treatment due to the possibility of obtaining higher delignification ranges, increase the porosity and the available surface area, and decrease the non-specific adsorption of hydrolytic enzymes [19,24,25]. Thus, when alkali-treated WIS were subsequently subjected to laccase treatment, a synergistic effect was observed in the saccharification process, enhancing sugar recovery yields by 21% ($p < 0.05$) and 30% ($p < 0.01$) in RGR and RXR, respectively (Figure 1). The increase in both porosity and surface area promoted by the mild alkali extraction enables an easier penetration of laccase into the fibers, allowing a better accessibility to the lignin polymer. Similar results were found by Yang *et al.* [25] when using *Brassica campestris* straw as raw material. These authors observed by scanning electron microscopy (SEM) some irregular holes on the surface of *B. campestris* straw after alkali treatment, being increased not only in number and density but also in width and depth when the laccase extracted from the fungus *Ganoderma lucidum* was subsequently used. The same effect was described by Li *et al.* [19] in corn straw after combining pretreatment with NaOH and crude ligninolytic enzyme produced by the fungus *Trametes hirsuta*. These results strongly highlight the benefits of combining a mild alkali treatment with a bacterial laccase treatment for improving the hydrolysability of steam-exploded wheat straw.

3.3. Laccase Treatment and Fermentation of the Whole Slurry

In addition to offering the possibility of increasing the sugar content during the enzymatic hydrolysis, laccase can work as a detoxification agent to improve the fermentability of pretreated lignocellulosic materials [7]. With the aim of evaluating the effect of bacterial laccase treatment on the fermentability of steam-pretreated wheat straw, the whole slurry was subjected to laccase treatment and simultaneous saccharification and fermentation without and with a presaccharification step: Strategy 3, consecutive laccase treatment and simultaneous saccharification and fermentation (LSSF); and Strategy 4, consecutive laccase treatment with presaccharification and simultaneous saccharification and fermentation (LPSSF).

3.3.1. Effect of Bacterial Laccase Treatment on Inhibitory Compounds

The concentration of inhibitory compounds after treatment with MetZyme laccase, without and with an enzymatic presaccharification step, was determined and compared with their respective controls assays (Table 3). Inhibitory compounds can alter the growth of fermenting microorganisms and also inhibit/deactivate cellulolytic enzymes, decreasing final yields and productivities [5,36–38]. Furfural and 5-HMF have a direct inhibition effect on either the glycolytic or fermentative enzymes of the yeast, reducing equally biomass formation and ethanol yields. Acetic acid and formic acid reduce biomass formation by modifying the intracelular pH and promoting an imbalance in the ATP/ADP ratio. Finally, phenols alter biological membranes, affecting the growth rates and also inhibiting and deactivating hydrolytic enzymes.

Table 3. Inhibitory compounds concentration (g/L) of slurry samples treated with bacterial MetZyme laccase without or with enzymatic presaccharification.

Assay	Inhibitors (g/L)				
	Furfural	5-HMF	Acetic Acid	Formic Acid	Phenols
C	0.3 ± 0.0	0.1 ± 0.0	2.5 ± 0.0	2.2 ± 0.1	1.6 ± 0.0
L	0.3 ± 0.0	0.1 ± 0.0	2.4 ± 0.0	2.3 ± 0.4	1.3 ± 0.0 **
CP	0.3 ± 0.0	0.1 ± 0.0	3.3 ± 0.0	2.1 ± 0.0	1.8 ± 0.1
LP	0.3 ± 0.0	0.1 ± 0.0	3.3 ± 0.0	2.1 ± 0.0	1.4 ± 0.1 ***

C, control without presaccharification; L, laccase treatment without presaccharification; CP, control with presaccharification; LP, laccase treatment with presaccharification; 5-HMF, 5-hydroxymethylfurfural. Analysis of variance (ANOVA) was performed to identify differences between C and L or CP and LP. Difference in means is significant at the (**) 0.01 or (***) 0.001 level.

In general, laccases catalyze the oxidation of phenols, generating unstable phenoxy radicals. These newly formed radicals interact with each other and lead to polymerization into aromatic compounds with lower inhibitory capacity. Total depletion in phenolic content seems to be impossible due to the structural characteristics of phenols [39]. Laccases can easily convert certain compounds, such as syringaldehyde or cinnamic acids, whilst other phenolic compounds are oxidized with lower rates (vanillin) or remain intact (hydroxybenzaldehyde) [31,39]. Several studies have reported an incomplete removal of phenolic compounds. Kalyani *et al.* [18] described a phenol reduction of 76% when the whole slurry from steam-exploded rice straw was treated with *Coltricia perennis* laccase. Moreno *et al.* [22] achieved higher phenol reductions (93%–94%) when *P. cinnabarinus* and *T. villosa* laccases were used on steam-exploded wheat straw. The same range was observed by Jurado *et al.* [17] with steam-exploded wheat straw and *Coriolopsis rigida* laccase and by Jönsson *et al.* [16] with SO_2 steam-pretreated willow and *Trametes versicolor* laccase. These mentioned studies have in common the use of fungal laccases, mainly from white-rot basidiomycetes. In our case, the bacterial MetZyme laccase decreased the total phenolic content by 20%–21% ($p < 0.01$), independently of whether a presaccharification step is included (Table 3). In comparison to fungal laccases, the lower efficiency in phenol removal by this particular bacterial laccase can be attributed to the lower redox potential of bacterial laccases in general. The redox potential of fungal laccases is estimated to be around +730 mV and +790 mV, while bacterial or plant laccases have a redox potential of about +450 mV. This higher redox potential of fungal laccases increases their capability to act towards a wider range of phenolic compounds. Nevertheless, the lower redox potential might not represent the only explanation for the reduction in the oxidation capacity of bacterial laccases, as other factors such as K_{cat}/K_M ratio (as a measure of enzyme efficiency) may also play an important role [40].

In contrast to the phenols reduction, furan derivatives and weak acids were altered by bacterial laccase in none of the strategies assayed (Table 3). The absence of laccase action on these type of inhibitory compounds has been already reported in previous studies with fungal laccases [16,17,22,31]. This substrate-specific reaction of laccases towards phenols offers some advantages over chemical and physical detoxification methods such as mild reaction conditions, the generation of fewer inhibitory sub-products and lower energy [41].

3.3.2. Effect of Bacterial Laccase Treatment on Cell Viability and Ethanol Production

Control and detoxified slurries, resulting from MetZyme laccase treatments without (L) and with the enzymatic presaccharification (LP), were subjected to SSF for 72 h at 42 °C. *K. marxianuns* CECT 10875 was used as the fermenting microorganism due to its ability to tolerate relatively high temperatures. Thermotolerant yeasts are gaining great significance due to the possibility of better integration between both saccharification and fermentation stages. Optimal temperatures for enzymatic hydrolysis are about 50 °C. In this context, the use of thermotolerant yeasts capable of growing and fermenting around those temperatures is beneficial for the performance of hydrolytic enzymes [31]. During fermentation assays, cell viability, glucose consumption and ethanol production were monitored (Figures 2 and 3).

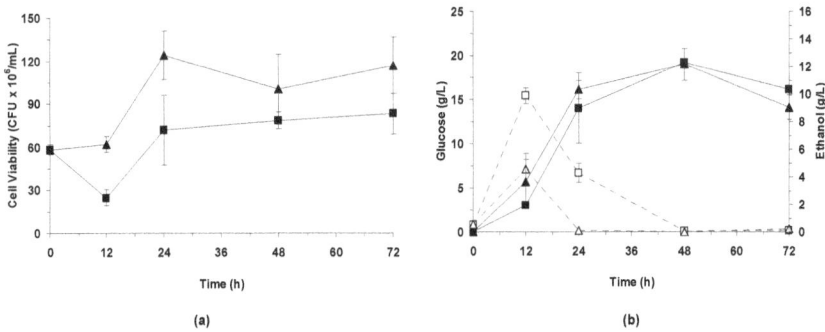

(a)   (b)

Figure 2. Consecutive laccase treatment and simultaneous saccharification and fermentation (LSSF, Strategy 3). (**a**) Viable cells during simultaneous saccharification and fermentation (SSF) assay with *K. marxianus* of slurry samples resulting from Metzyme laccase treatment. Symbols used: control (■) and laccase (▲) samples. (**b**) Time course for ethanol (filled symbols and continuous lines) and glucose (open symbols and discontinuous lines) during simultaneous saccharification and fermentation (SSF) assay with *K. marxianus* of slurry samples resulting from Metzyme laccase treatment. Symbols used: control (■, □) and laccase (▲, △) samples. Mean values and standard deviations were calculated from the triplicates to present the results.

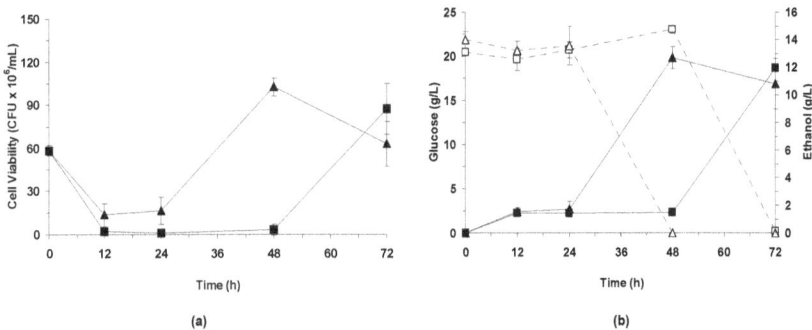

(a)   (b)

Figure 3. Consecutive laccase treatment with presaccharification and simultaneous saccharification and fermentation (LPSSF, Strategy 4). (**a**) Viable cells during simultaneous saccharification and fermentation (SSF) assay with *K. marxianus* of slurry samples resulting from Metzyme laccase treatment with presaccharification. Symbols used: control (■) and laccase (▲) samples. (**b**) Time course for ethanol (filled symbols and continuous lines) and glucose (open symbols and discontinuous lines) during simultaneous saccharification and fermentation (SSF) assay with *K. marxianus* of slurry samples resulting from Metzyme laccase treatment with presaccharification. Symbols used: control (■, □) and laccase (▲, △) samples. Mean values and standard deviations were calculated from the triplicates to present the results.

LSSF

During SSF of wheat straw slurry, cell viability—in the form of CFU/mL—decreased within the first 12 h in control assays (Figure 2A). This effect is correlated with the adaptation of the yeast to the different components in the fermentation medium, and usually promotes a delay in glucose consumption and ethanol production (Figure 2B). This adaptation period depends on different factors, such as the presence of inhibitory compounds, the nature and concentration of inhibitors and the synergistic effects between them [5,36].

The adaptation phase is overcome by *K. marxianus* after converting certain inhibitory compounds, including furfural and 5-HMF, to their less toxic forms. After being adapted, the yeast showed a remarkable increase in cell viability between 12 and 24 h of SSF, until reaching the value of 80 CFU/mL that remained constant for the rest of the process (Figure 2A). The increase in cell viability made it possible to obtain the maximum glucose consumption (values were not estimated due to the continuous release of glucose) and ethanol production rates (0.59 g/L· h between 12 and 24 h), lowering the glucose concentration to values below 0.1 g/L and reaching the highest ethanol concentration (12.3 g/L) within 48 h (Figure 2B, Table 4).

Table 4. Summary of simultaneous saccharification and fermentation (SSF) assays with *K. marxianus* of slurry samples treated with bacterial MetZyme laccase without or with enzymatic presaccharification.

Assay	Adaptation Phase (h)	EtOH$_{max}$ (g/L)	$Y_{E/G}$ (g/g)	$Y_{E/ET}$ (%)	Q_{Emax} (g/L· h)
C	12	12.3 ± 0.1 [a]	0.29 ± 0.00	65.1	0.58 ± 0.21 [c]
L	<12	12.2 ± 1.2 [a]	0.29 ± 0.03	64.4	0.56 ± 0.09 [c]
CP	48	12.0 ± 0.1 [b]	0.29 ± 0.1	63.4	0.43 ± 0.00 [d]
LP	24	12.7 ± 0.8 [a]	0.30 ± 0.1	67.1	0.44 ± 0.02 [e]

C, control without presaccharification; L, laccase treatment without presaccharification; CP, control with presaccharification; LP, laccase treatment with presaccharification; EtOH$_{max}$, maximum ethanol concentration reached at 48 h [a] and 72 h [b]; $Y_{E/G}$, ethanol yield based on total glucose content in the slurry. The ethanol yield is calculated considering that the liquid volume of the SSF system is constant [42]; $Y_{E/ET}$, percentage of ethanol produced from potential glucose, assuming maximum ethanol yields 0.45 g/g in *K. marxianus* [43]; Q_{Emax}, volumetric ethanol productivity within 12–24 h [c], 48–72 h [d] or 24–48 h [e] of SSF.

When laccase-treated slurries were subjected to SSF, the fermentation performance of *K. marxianus* was slightly improved due to the lower phenolic content. This improvement was more evident in cell viability, where no reduction in the number of CFU/mL was observed within the first 12 h and a significant increase to about 120 CFU/mL was obtained between 24 and 72 h of fermentation (Figure 2A). Similar values of maximum ethanol production rates and maximum ethanol concentrations were observed for both control and laccase-treated slurries (Figure 2B, Table 4). However, a shorter adaptation phase was observed in those slurries treated with Metzyme, which can aid in reducing the overall process time. Similar improvements on the fermentation performance of *K. marxianus* and other fermenting microorganisms have been also observed when using fungal laccases. Moreno *et al.* [22,31] reported higher cell viability, glucose consumption rates and ethanol productivity values when steam-exploded wheat straw was treated with *P. cinnabarinus* and *T. villosa* laccases. Jönsson *et al.* [16] reported higher glucose consumption rate, ethanol productivity and ethanol yield when the liquid fraction from acid steam-exploded willow was treated with *T. versicolor* laccase and fermented with *Saccharomyces cerevisiae*. In the same way, Jurado *et al.* [17] observed higher biomass concentration, sugar consumption and ethanol yield after treating steam-exploded wheat straw with *C. rigida* laccase and fermenting it with *S. cerevisiae*.

LPSSF

In comparison to assays without a presaccharification step, the enzymatic prehydrolysis (P) extended the adaptation period of yeast cells during the subsequent SSF process (Figure 3A,B). A remarkable drop in cell viability was measured in non-treated slurries within the first 12 h of SSF, followed by a long stationary phase where no cell growth was observed. After 48 h, a sudden increase in cell viability could be seen, reaching a value of about 95 CFU/mL (Figure 3A). Regarding glucose concentration, the prehydrolysis stage increased the glucose content before inoculation up to 20 g/L. After inoculation, the adaptation phase of *K. marxianus* allowed the continuous accumulation of this sugar during the first 48 h of SSF, reaching a maximum value of 23 g/L. Supported by the increase in cell viability, glucose started to be consumed after 48 h, and values below 0.1 g/L were observed at 72 h of SSF (Figure 3B). Maximum ethanol concentration (12 g/L) and yield (0.29 g/g) were similar to those obtained in SSF processes without presaccharification, but this value was reached with a delay of 24 h (Table 4). The extended lag phase of *K. marxianus* during PSSF processes in comparison to that observed during SSF can be justified by the presence of a higher concentration of inhibitory compounds after the presaccharification step, especially for acetic acid and phenols (Table 3). According to Thomsen *et al.* [44], acetic acid is produced by the hydrolysis of acetyl groups in hemicelluloses, which involves the synergistic action of both hemicellulase and acetyl esterase activities. In this sense, the cellulolytic NS50013 preparation used in this study—obtained from *Trichoderma* spp. strains—contains xylanase and acetyl esterase activities that can release the acetyl groups that remain in the fibers, increasing the acetic acid concentration [45]. Similarly, the increment in the phenol content can be explained by the release of *p*-coumaric acid and ferulic acid during the presaccharification step. Ferulic acid and *p*-coumaric acid are two typical lignin phenolic compounds present in wheat straw [3,31]. The hydrolysis of these cinnamic acids is attributed to the complementary action of xylanase and phenolic acid esterase activities [31]. The esterase activities, highlighting feruloyl esterase activity, are naturally produced by *Aspergillus niger*, which is the strain producing the β-glucosidase NS50010 preparation [46].

When laccase treatment was combined with the presaccharification step (LP), the adaptation phase was reduced from 48 h in control assays to 24 h in laccase-treated slurries. This reduction can be seen in Figure 3A, where an increase from about 15 CFU/mL to above 100 CFU/mL was observed within 24 and 48 h of SSF. In comparison to LSSF process without a presaccharification stage where the number of CFU/mL was kept constant after the adaptation phase, in LPSSF process a remarkable decrease in cell viability took place within 48 and 72 h, which is an indicator of the higher inhibitory content even after laccase treatment. In relation to glucose consumption and ethanol production, similar rates (0.43 g/L·h and 0.44 g/L·h in control and laccase-treated slurries, respectively) were observed after the adaptation phase either in control or laccase-treated assays, which resulted in a maximum ethanol concentration of about 12 g/L—similar to those values obtained in the SSF processes without a presaccharification step (Figure 3B, Table 4). It is important to notice that higher ethanol concentrations than those obtained in the present work are needed for the cost-effectiveness of a commercial lignocellulosic ethanol production. Working at higher substrate concentrations is therefore imperative and a presaccharification step is typically included in the process in order to avoid certain problems such as mixing. In this context, the use of laccases could play a crucial role to increase the fermentability of steam-pretreated lignocellulosic materials.

4. Conclusions

The present work shows the potential of the bacterial MetZyme laccase for improving both the hydrolysability and fermentability of steam-pretreated materials. The laccase treatment of the WIS fraction resulted in slightly higher glucose and xylose recoveries during a saccharification process. This improvement was increased synergistically by the action of a mild alkaline extraction performed prior to laccase treatment. MetZyme laccase also showed modest phenol removal when treating the whole pretreated slurries, reducing the inhibitory effects of steam-exploded wheat straw. The lower inhibitory content led to improve the fermentation performance of *K. marxianus* in SSF processes with or without a presaccharification step, shortening its adaptation period and the overall fermentation times. These results represent an interesting approach to improve the efficiency of the ethanol production process, which might contribute to making lignocellulosic ethanol production economically viable. Nevertheless, other parameters, including laccase dosages and production costs, need to be further studied and optimized for the cost-effectiveness of the process.

Acknowledgments: Authors wish to thank the Spanish MIMECO for funding this study via Project CTQ2013-47158-R. Antonio D. Moreno acknowledges a "Juan de la Cierva" contract (FJCI-2014-22385).

Author Contributions: Antonio D. Moreno, David Ibarra, Antoine Mialon, and Mercedes Ballesteros participated in the design of the study. Antonio D. Moreno and David Ibarra performed the experimental work and wrote the manuscript. Antonio D. Moreno, David Ibarra, Antoine Mialon, and Mercedes Ballesteros conceived the study and commented on the manuscript. All the authors read and approved the final manuscript.

Conflicts of Interest: Antoine Mialon is Application Team Leader at Metgen Oy.

Abbreviations

The following abbreviations are used in this manuscript:

WIS	Water Insoluble Solids fraction
NREL-LAP	National Renewable Energies Laboratory-Laboratory Analytical Procedures
SSF	Simultaneous Saccharification and Fermentation
CFU	Colony Forming Units
DW	Dry Weight
LAP	Laboratory Analytical Procedures
IU	International Units
FPU	Filter Paper Units
TS	Total Solids
RGR	Relative Glucose Recovery
RXR	Relative Xylose Recovery
LSSF	consecutive Laccase treatment and Simultaneous Saccharification and Fermentation
LPSSF	consecutive Laccase treatment with Presaccharification and Simultaneous Saccharification and Fermentation
5-HMF	5-hydroxymethylfurfural
C	Control treatment
CP	Control treatment with Presaccharification
L	Laccase treatment
LP	Laccase treatment with Presaccharification
Alk	Alkaline extraction
ATP/ADP	Adenosine-5′-triphosphate/Adenosine-5′-diphosphate

References

1. Ballesteros, M. Enzymatic hydrolysis of lignocellulosic biomass. In *Bioalcohol Production. Biochemical Conversion of Lignocellulosic Biomass*; Waldron, K., Ed.; Woodhead Publishing: Cambridge, UK, 2010; pp. 159–177.
2. Alvira, P.; Tomás-Pejó, E.; Ballesteros, M.; Negro, M.J. Pretreatment technologies for an efficient bioethanol production process based on enzymatic hydrolysis: A review. *Bioresour. Technol.* **2010**, *101*, 4851–4861. [CrossRef] [PubMed]
3. Ballesteros, I.; Negro, M.; Oliva, J.M.; Manzanares, P.; Ballesteros, M. Ethanol production from steam-explosion pretreated wheat straw. *Appl. Biochem. Biotechnol.* **2006**, *130*, 496–508. [CrossRef]
4. Berlin, A.; Balakshin, M.; Gilkes, N.; Kadla, J.; Maximenko, V.; Kubo, S. Inhibition of cellulase, xylanase and β-glucosidase activities by sofwood lignin preparations. *J. Biotechnol.* **2006**, *125*, 198–209. [CrossRef] [PubMed]
5. Palmqvist, E.; Hahn-Hägerdal, B. Fermentation of lignocellulosic hydrolysates. II: Inhibitors and mechanism of inhibition. *Bioresour. Technol.* **2000**, *74*, 25–33. [CrossRef]
6. Palmqvist, E.; Hahn-Hägerdal, B. Fermentation of lignocellulosic hydrolysates. I: Inhibition and detoxification. *Bioresour. Technol.* **2000**, *74*, 17–24. [CrossRef]
7. Moreno, A.D.; Ibarra, D.; Alvira, P.; Tomás-Pejó, E.; Ballesteros, M. A review of biological delignification and detoxification methods for lignocellulosic bioethanol production. *Crit. Rev. Biotechnol.* **2015**, *35*, 342–354. [CrossRef] [PubMed]
8. Thurston, C.F. The structure and function of fungal laccases. *Microbiology* **1994**, *140*, 19–26. [CrossRef]

9. Enguita, F.J.; Matias, P.M.; Martins, L.O.; Placido, D.; Henriques, A.O.; Carrondo, M.A. Spore-coat laccase CotA from *Bacillus subtilis*: Crystallization and preliminary X-ray characterization by the MAD method. *Acta Crystallogr. D Biol. Cristallogr.* **2002**, *58*, 1490–1493. [CrossRef]
10. Bajpai, P. Biological bleaching of chemical pulps. *Crit. Rev. Biotechnol.* **2004**, *24*, 1–58. [CrossRef] [PubMed]
11. Paice, M.G.; Bourbonnais, R.; Reid, I.D.; Archibald, F.S.; Jurasek, L. Oxidative bleaching enzymes: A review. *J. Pulp Pap. Sci.* **1995**, *21*, 280–284.
12. Ibarra, D.; Camarero, S.; Romero, J.; Martínez, M.J.; Martínez, A.T. Integrating laccase-mediator treatment into an industrial-type sequence for totally chlorine free bleaching of eucalipt kraft pulp. *J. Chem. Technol. Biotechnol.* **2006**, *81*, 1159–1165. [CrossRef]
13. Eugenio, M.E.; Hernández, M.; Moya, R.; Martín-Sampedro, R.; Villar, J.C.; Arias, M.E. Evaluation of a new laccase produced by Streptomyces ipomoea on biobleaching and ageing of kraft pulps. *BioResources* **2011**, *6*, 3231–3241.
14. Chandra, R.; Chowdhary, P. Properties of bacterial laccases and their application in biorremediation of industrial wastes. *Environ. Sci. Process. Impacts* **2015**, *17*, 326–342. [CrossRef] [PubMed]
15. Milstein, O.; Haars, A.; Majcherczyk, A.; Tautz, D.; Zanker, H.; Hüttermann, A. Removal of chlorophenols and chlorolignins from bleaching effluents by combined chemical and biological treatment. *Water Sci. Technol.* **1988**, *20*, 161–170.
16. Jönsson, L.J.; Palmqvist, E.; Nilvebrant, N.-O.; Hahn-Hägerdal, B. Detoxification of wood hydrolysates with laccase and peroxidase from the white-rot fungus *Trametes versicolor*. *Appl. Microbiol. Biotechnol.* **1998**, *49*, 691–697. [CrossRef]
17. Jurado, M.; Prieto, A.; Martínez-Alcalá, A.; Martínez, A.T.; Martínez, M.J. Laccase detoxification of steam-exploded wheat straw for second generation bioethanol. *Bioresour. Technol.* **2012**, *100*, 6378–6384. [CrossRef] [PubMed]
18. Kalyani, D.; Dhiman, S.S.; Kim, H.; Jeya, M.; Kim, I.-W.; Lee, J.-K. Characterization of a novel laccase from the isolated *Coltricia perennis* and its application to detoxification of biomass. *Process Biochem.* **2012**, *47*, 671–678. [CrossRef]
19. Li, J.; Sun, F.; Li, X.; Yan, Z.; Yuan, Y.; Liu, X.F. Enhanced saccharification of corn straw pretreated by alkali combining crude ligninolytic enzymes. *J. Chem. Technol. Biotecnhol.* **2012**, *87*, 1687–1693. [CrossRef]
20. Martín-Sampedro, R.; Eugenio, M.E.; García, J.C.; López, F.; Villar, J.C.; Díaz, M.J. Steam explosion and enzymatic pre-treatments as an approach to improve the enzymatic hydrolysis of *Eucalytpus globulus*. *Biomass Bioenergy* **2012**, *42*, 97–106. [CrossRef]
21. Moilanen, U.; Kellock, M.; Galkin, S.; Viikari, L. The laccase-catalyzed modification of lignin for enzymatic hydrolysis. *Enzym. Microb. Technol.* **2011**, *49*, 492–498. [CrossRef] [PubMed]
22. Moreno, A.D.; Ibarra, D.; Fernández, J.L.; Ballesteros, M. Different laccase detoxification strategies for ethanol production from lignocellulosic biomass by the thermotolerant yeast *Kluyveromyces marxianus* CECT 10875. *Bioresour. Technol.* **2012**, *106*, 101–109. [CrossRef] [PubMed]
23. Moreno, A.D.; Ibarra, D.; Alvira, P.; Tomás-Pejó, E.; Ballesteros, M. Exploring laccase and mediators behavior during saccharification and fermentation of steam-exploded wheat straw for bioethanol production. *J. Chem. Technol. Biotehcnol.* **2015**. [CrossRef]
24. Palonen, H.; Viikari, L. Role of oxidative enzymatic treatments on enzymatic hydrolysis of softwood. *Biotechnol. Bioeng.* **2004**, *86*, 550–557. [CrossRef] [PubMed]
25. Yang, P.; Jiang, S.; Zheng, Z.; Shuizhong, L.; Luo, S.; Pan, L. Effect of alkali and laccase pretreatment of Brassica campestris straw: Architecture, crystallisation, and saccharification. *Polym. Renew. Resour.* **2011**, *2*, 21–34.
26. National Renewable Energy Laboratory. *Chemical Analysis and Testing Laboratory Analytical Procedures*; National Renewable Energy Laboratory: Golden, CO, USA, 2007. Available online: www.1.ere.energy.gov/biomass/analytical_procedures.html (accessed on 15 March 2015).
27. Ghose, T.K. Measurement of cellulase activity. *Pure Appl. Chem.* **1987**, *59*, 257–268. [CrossRef]
28. Ballesteros, I.; Ballesteros, M.; Cabañas, A.; Carrasco, J.; Martín, C.; Negro, M.J.; Saez, F.; Saez, R. Selection of thermotolerant yeasts for simultaneous saccharification and fermentation (SSF) of cellulose to ethanol. *Appl. Biochem. Biotechnol.* **1991**, *28*, 307–315. [CrossRef] [PubMed]
29. Singleton, V.L.; Rossi, J.A. Colorimetry of total phenolic with phosphomolibdic-phosphotungstic acid reagent. *Am. J. Enol. Vitic.* **1965**, *16*, 144–158.

30. Oliva, J.M.; Saez, F.; Ballesteros, I.; González, A.; Negro, M.J.; Manzanares, P.; Ballesteros, M. Effect of lignocellulosic degradation compounds from steam explosion pretreatment on ethanol fermentation by thermotolerant yeast *Kluyveromyces marxianus. Appl. Biochem. Biotechnol.* **2003**, *105*, 141–154. [CrossRef]

31. Moreno, A.D.; Ibarra, D.; Ballesteros, I.; Fernández, J.L.; Ballesteros, M. Ethanol from laccase-detoxified lignocellulose by the thermotolerant yeast *Kluyveromyces marxianus*—Effects of steam pretreatment conditions, process configurations and substrate loadings. *Biochem. Eng. J.* **2013**, *79*, 94–103. [CrossRef]

32. Martín-Sampedro, R.; Capanema, E.A.; Hoeger, I.; Villar, J.C.; Rojas, O.J. Lignin changes after steam explosion and laccase-mediator treatment of eucalytpus wood chips. *J. Agric. Food Chem.* **2011**, *59*, 8761–8769. [CrossRef] [PubMed]

33. Oliva-Taravilla, A.; Moreno, A.D.; Demuez, M.; Ibarra, D.; Tomás-Pejó, E.; González-Fernández, C.; Ballesteros, M. Unraveling the effects of laccase treatment on enzymatic hydrolysis of steam-exploded wheat straw. *Bioresour. Technol.* **2015**, *175*, 209–215. [CrossRef] [PubMed]

34. Fernandes, M.C.; Ferro, M.D.; Paulino, A.F.C.; Mendes, J.A.S.; Gravitis, J.; Evtuguin, D.V.; Xavier, A.M.R.B. Enzymatic saccharification and bioethanol production from *Cynara cardunculus* pretreated by steam explosion. *Bioresour. Technol.* **2015**, *186*, 309–315. [CrossRef] [PubMed]

35. Yang, B.; Boussaid, A.; Mansfield, S.D.; Gregg, D.J.; Sadler, J.N. Fast and efficient alkaline peroxide treatment to enhance the enzymatic digestibility of steam-exploded softwood substrates. *Biotechnol. Bioeng.* **2002**, *77*, 678–684. [CrossRef] [PubMed]

36. Klinke, H.B.; Thomsen, A.B.; Ahring, B.K. Inhibition of ethanol-producing yeast and bacteria by degradation products produced during pretreatment of biomass. *Appl. Microbiol. Biotechnol.* **2004**, *66*, 10–26. [CrossRef] [PubMed]

37. Ximenes, E.; Kim, Y.; Mosier, N.; Dien, B.; Ladisch, M. Inhibition of cellulases by phenols. *Enzym. Microb. Technol.* **2010**, *46*, 170–176. [CrossRef]

38. Ximenes, E.; Kim, Y.; Mosier, N.; Dien, B.; Ladisch, M. Deactivation of cellulases by phenols. *Enzym. Microb. Technol.* **2011**, *48*, 54–60. [CrossRef] [PubMed]

39. Kolb, M.; Sieber, V.; Amann, M.; Faulstich, M.; Schieder, M. Removal of monomer delignification products by laccase from *Trametes versicolor. Bioresour. Technol.* **2012**, *104*, 298–304. [CrossRef] [PubMed]

40. Lahtinen, M.; Kruus, K.; Boer, H.; Kemell, M.; Andberg, M.; Viikari, L.; Sipilä, J. The effect of lignin model compound structure on the rate of oxidation catalyzed by two different fungal laccases. *J. Mol. Catal. B Enzym.* **2009**, *57*, 204–210. [CrossRef]

41. Parawira, W.; Tekere, M. Biotechnological strategies to overcome inhibitors in lignocellulose hydrolysates for ethanol production: Review. *Crit. Rev. Biotechnol.* **2011**, *31*, 20–31. [CrossRef] [PubMed]

42. Zhang, J.; Bao, J. A modified method for calculating practical ethanol yield at high lignocellulosic solids content and high ethanol titer. *Bioresour. Technol.* **2012**, *116*, 74–79. [CrossRef] [PubMed]

43. Tomás-Pejó, E.; Oliva, J.M.; González, A.; Ballesteros, I.; Ballesteros, M. Bioethanol production from wheat straw by the thermotolerant yeast *Kluyveromyces marxianus* CECT 10875 in a simultaneous saccharification and fermentation fed-batch process. *Fuel* **2009**, *88*, 2142–2147. [CrossRef]

44. Thomsen, M.T.; Thygesen, A.; Thomsen, A.B. Identification and characterization of fermentation inhibitors formed during hydrothermal treatment and following SSF of wheat straw. *Appl. Microbiol. Biotechnol.* **2009**, *83*, 447–455. [CrossRef] [PubMed]

45. Juhász, T.; Szengyel, K.; Réczey, K.; Siika-Aho, M.; Viikari, L. Characterization of cellulase and hemicellulases produced by *Trichoderma reesei* on various carbon sources. *Process Biochem.* **2005**, *40*, 3519–3525. [CrossRef]

46. Dien, B.S.; Ximenes, E.A.; O'Brien, P.J.; Moniruzzaman, M.; Li, X.L.; Balan, V.; Dale, B.; Cotta, M.A. Enzyme characterization for hydrolysis of AFEX and liquid hot water pretreated distillers'grains and their conversion to ethanol. *Bioresour. Technol.* **2008**, *99*, 5216–5225. [CrossRef] [PubMed]

fermentation   MDPI

Review
Application of Non-*Saccharomyces* Yeasts to Wine-Making Process

José Juan Mateo * and Sergi Maicas

Departament de Microbiologia i Ecologia, Universitat de València, E-46100-Burjassot, Spain; Sergi.Maicas@uv.es
* Correspondence: Jose.J.Mateo@uv.es; Tel.: +34-96-3543008

Academic Editor: Ronnie G. Willaert
Received: 18 May 2016; Accepted: 16 June 2016; Published: 23 June 2016

Abstract: Winemaking is a complex process involving the interaction of different microbes. The two main groups of microorganisms involved are yeasts and bacteria. Non-*Saccharomyces* yeasts are present on the grape surface and also on the cellar. Although these yeasts can produce spoilage, these microorganisms could also possess many interesting technological properties which could be exploited in food processing. It has been shown that some of the metabolites that these yeasts produce may be beneficial and contribute to the complexity of the wine and secrete enzymes providing interesting wine organoleptic characteristics. On the other hand, non-*Saccharomyces* yeasts are the key to obtain wines with reduced ethanol content. Among secreted enzymes, β-glucosidase activity is involved in the release of terpenes to wine, thus contributing to varietal aroma while β-xylosidase enzyme is also interesting in industry due to its involvement in the degradation of hemicellulose by hydrolyzing its main heteroglycan (xylan).

Keywords: non-*Saccharomyces* yeasts; wine; flavor; β-glucosidase; β-xylosidase

---

1. Introduction

Since Louis Pasteur elucidated the conversion of grape juice into wine, this process and the role of the yeast therein has been studied extensively [1]. More than 130 years later, there are many areas that are still not well understood [2]. This is especially the case for the roles of the numerous non-*Saccharomyces* yeasts normally associated with grape must and wine. These yeasts, present in all wine fermentations, are metabolically active and their metabolites can impact on wine quality. In the past, the influence of non-*Saccharomyces* yeasts in wine was restricted and even eliminated by inoculation with pure *S. cerevisiae* cultures because they have been regarded as spoilage yeasts [3,4]. However, in the past three decades, great interest has grown in the beneficial role of non-*Saccharomyces* yeasts in wine biotechnology [5,6]. It has been shown that some of the metabolites that these yeasts produce may be beneficial and contribute to the complexity of the wine when they are used in mixed fermentations with *S. cerevisiae* cultures [7,8]. Evidence supporting this fact has been published [9] and the role of the non-*Saccharomyces* yeasts in wine fermentation is receiving increasingly more attention by wine microbiologists in wine-producing countries [10].

Non-*Saccharomyces* yeasts are found on the grapes, but also in lesser numbers on the cellar equipment [11]. The initial environment that affects the microbial makeup of a wine fermentation is that of the vineyard. Although a drastically different environment than juice or wine, the types of microbes present on grapes will have an impact on the ensuing ecology in the wine fermentation, particularly in the early stages. Microorganisms appear to colonize around the grape stomata where small amounts of exudate are secreted [12,13]. The apiculate yeasts, *Hanseniaspora* and *Kloeckera*, its asexual anamorph, are the most prevalent vineyard yeasts and typically represent over half the yeast flora on grapes [14]. Other yeast genera present on berries include: *Metschnikowia, Candida,*

Pichia, Wickerhamomyces, Zygosaccharomyces, and *Torulaspora* [15]. Also present in the vineyard are numerous other yeasts, some of which have an impact on wine: *Sporidiobolus, Kluyveromyces,* and *Hansenula* [16]. *Saccharomyces* species are relatively scarce among healthy berries (Table 1) [17,18]. Before inoculation with *S. cerevisiae,* they are the yeasts present in the highest numbers in the grape must. During the fermentation there is a sequence of dominance by the various non-*Saccharomyces* yeasts, followed by *S. cerevisiae,* which then completes the fermentation [19]. This is especially evident in spontaneously fermenting grape must, which has a low initial *S. cerevisiae* concentration. Research has shown that non-*Saccharomyces* yeast strains can be detected throughout wine fermentation [20] and their dominance during the early part of fermentation can leave an imprint on the final composition of the wine [21].

Table 1. Main non-*Saccharomyces* yeasts isolated from grape musts and wines.

Aureobasidium pullulans	*Hansenula* sp
Brettanomyces sp	*Issatchenkia terricola*
B. anomalus	*Kluyveromyces thermotolerans*
Candida guilliermondii	*Lachancea thermotolerans*
C. molischiana	*Metschnikowia pulcherrima/C. pulcherrima*
C. stellata	*Pichia angusta*
C. utilis	*P. anomala*
C. zemplinina	*P. capsulata*
Debaryomyces castellii	*P. guilliermondii*
D.hansenii	*P. kluyvery*
D.polymorphus	*P. membranifaciens*
D.pseudopolymorphus	*Saccharomycodes ludwigii*
D. vanriji	*Schizosaccharomyces pombe*
Hanseniaspora sp. (*Kloeckera*)	*Sporidiobolus pararoseus*
H. guilliermondii	*Torulaspora delbrueckii*
H. osmophila	*Trichosporon asahii*
H. vineae	*Wickerhamomyces anomalus*
H. uvarum	*Zygosaccharomyces bailii*

In the past, the influence of non-*Saccharomyces* yeasts in wine was restricted and even eliminated by inoculation with pure *S. cerevisiae* cultures because they have been regarded as spoilage yeasts [17]. However, in the past three decades, great interest has grown in the beneficial role of non-*Saccharomyces* yeasts in wine biotechnology [18,19]. It has been shown that some of the metabolites that these yeasts produce may be beneficial and contribute to the complexity of the wine when they are used in mixed fermentations with *S. cerevisiae* cultures [20,21].

It is believed that when pure non-*Saccharomyces* yeasts are cultivated with *S. cerevisiae* strains, their negative metabolic activities may not be expressed or could be modified by the metabolic activities of the *S. cerevisiae* strains [22]. Several strains belonging to different non-*Saccharomyces* species have been extensively studied in relation to the formation of some metabolic compounds affecting the bouquet of the final product. Diverse studies on the growth and metabolic interactions between non-*Saccharomyces* and *Saccharomyces* yeasts in mixed cultures have shown that their impact on ethanol content, wine flavor, aromatic profile, and quality and control of spoilage yeasts depends on the strains and the inoculation strategies [23,24]. In addition, a great number of studies inform about enzyme activities in winemaking and fermentations [25,26]. However, there are no known reports that associate the production of enzymatic activities in mixed cultures of *Saccharomyces* and non-*Saccharomyces* during the fermentation with the final aromatic profile of wines.

2. Contribution of Non-*Saccharomyces* Yeast Reduction in the Ethanol Content of Wines

Consumer and market demand for wines containing lower ethanol has shaped research to develop and evaluate strategies to generate low-ethanol wines [27]. Numerous studies have reported lower ethanol yields when using non-*Saccharomyces* yeast [23,28]. Another alternative is to exploit the oxidative metabolism observed in some non-*Saccharomyces* species [29]. Nevertheless, only one study has reported the use of aerobic yeast for the production of reduced alcohol wine. Wines containing 3% *v/v* ethanol were obtained after fermentation of grape must by *Williopsis saturnus* and *Pichia subpelliculosa* under intensive aerobic conditions. These reduced alcohol wines were considered to be of an adequate quality [30].

Microbiological approaches for decreasing ethanol concentrations take advantage of the differences in energy metabolism among the wine yeast species. Several strategies that use genetically-modified yeasts have been proposed for the production of low-alcohol wine. Recently, Tilloy and co-workers [31] using evolution-based strategies together with breeding strategy showed that evolved or hybrid strains produced an ethanol reduction of 0.6%–1.3% (v/v). Another approach to reduce the production of ethanol could be the use of non-*Saccharomyces* wine yeasts, in combination with *S. cerevisiae*, to improve the quality and enhance the complexity of wine. Following numerous studies on the influence of non-*Saccharomyces* yeast in winemaking, there has been a re-evaluation of the role of these yeasts. Indeed, some non-*Saccharomyces* yeast can enhance the profile of the wine, and for this reason the use of controlled multi-starter fermentation using selected cultures of non-*Saccharomyces* and *S. cerevisiae* yeast strains has been encouraged [32]. Indeed, nowadays one of the most recent technological advances in winemaking is the practice of co-inoculation of grape juice with selected culture of a non-*Saccharomyces* coupled with a *S. cerevisiae* starter strain [25]. In this context, non-*Saccharomyces* wine yeasts in multi-starter fermentations could be an interesting way to reduce the ethanol content in wine. In addition, different respiro-fermentative regulatory mechanisms of some non-*Saccharomyces* yeasts compared to *S. cerevisiae* could be a modality to reduce the ethanol production through partial and controlled aeration of the grape juice. Indeed, in this way sugar is consumed via respiration rather than fermentation. Both of these approaches have indicated the promising use of non-*Saccharomyces* wine yeast to limit ethanol production [33,34].

3. Contribution to Wine by Non-*Saccharomyces* Yeast

According to Fleet [35], yeast influences wine aroma by different mechanisms; of these, novo biosynthesis of aroma compounds is probably the most important [36]. The variety of odor compounds produced by non-*Saccharomyces* yeasts is known [37]. The contribution of non-*Saccharomyces* yeasts to wine quality can take various forms. Production of glycerol by *Candida stellata* and esters by *C. pulcherrima* has been reported [15]. Other non-*Saccharomyces* yeasts are also widely recognized for producing glucosidase enzymes [38], which, by hydrolyzing such bonds, are capable of releasing volatile compounds linked to sugars, giving greater complexity to the wine's aromatic profile [39]. Conversely, others such as *Kloeckera apiculata* are associated with the production of acetic acid, which lowers wine quality [40]. Therefore, to determine the potential of non-*Saccharomyces* yeasts to be used in the wine industry, it is necessary to check that their activity in mixed culture does not affect the development of *Saccharomyces*, or produce compounds that may harm wine quality. These metabolic products include terpenoids, esters, higher alcohols, glycerol, acetaldehyde, acetic acid, and succinic acid [41].

Therefore, sensory differences were found [42]. Over 160 esters have been distinguished in wine [43]. These esters can have a helpful effect on wine quality, especially in wine from varieties with neutral flavors [44]. Non-*Saccharomyces* can be divided into two groups, neutral yeasts (producing little or no flavor compounds) and flavor-producing species. Flavor-producing yeasts included *P. anomala* (*Hansenula anomala*) and *K. apiculata*. *Candida pulcherrima* is also known to be a high producer of esters [41]. The accumulation of esters in wine is determined by the balance between the yeast's ester-synthesizing enzymes and esterases (responsible for cleavage and in some cases, formation of ester bonds) [45]. Although extracellular esterases are known to occur in *S. cerevisiae* [46], the situation for non-*Saccharomyces* needs further investigation.

Different non-*Saccharomyces* yeasts produce different levels of higher alcohols (n-propanol, isobutanol, isoamyl alcohol, and active amyl alcohol) [44]. This is important during wine production, as high concentrations of higher alcohols are generally not desired, whereas lower values can add to wine complexity.

Glycerol, the next major yeast metabolite produced during wine fermentation after ethanol, is important in yeast metabolism for regulating redox potential in the cell [47]. Glycerol contributes to smoothness, sweetness, and complexity in wines, but the grape variety and wine style will govern the extent to which glycerol impacts on these properties [48]. Several non-*Saccharomyces* yeasts, particularly *C. zemplinina*, can consistently produce high glycerol concentrations during wine fermentation [25]. Unfortunately, increased glycerol production is usually linked to increased acetic acid production [49], which can be detrimental to wine quality. Spontaneously-fermented wines have higher glycerol levels, indicating a possible contribution by non-*Saccharomyces* yeasts [14].

Other compounds that are known to play a role in the sensory quality of wine include volatile fatty acids, carbonyl, and sulfur compounds [44]. Volatile thiols greatly contribute to the varietal character of some grape varieties, particularly Sauvignon Blanc [50]. Some non-*Saccharomyces* strains, specifically isolates from *C. zemplinina* and *P. kluyveri*, can produce significant amounts of the volatile thiols 3-mercaptohexan-1-ol (3MH) and 3-mercaptohexan-1-ol acetate (3MHA), respectively, in Sauvignon Blanc wines [51]. Similarly, *T. delbrueckii*, *M. pulcherrima*, and *Lachancea thermotolerans* have also been described as able to release important quantities of 3MH from its precursor during Sauvignon Blanc fermentation [52].

However, the use of some non-*Saccharomyces* yeast in mixed fermentations with *S. cerevisiae* can generate wines with increased volatile acidity and acetic acid concentration [25]. Some non-*Saccharomyces* yeasts are able to form succinic acid [48]. This correlates with high ethanol production and ethanol tolerance. Succinic acid production could positively influence the analytical profile of wines by contributing to the total acidity in wines with insufficient acidity. Nevertheless, succinic acid has a "salt-bitter-acid" taste and excessive levels will negatively influence wine quality. Other non-*Saccharomyces* metabolites can act as intermediaries in aroma metabolic pathways. Acetoin is considered a relatively odorless compound in wine [53]. However, diacetyl and 2,3-butanediol (potentially off-flavors in wine) can be derived from acetoin by chemical oxidation and yeast-mediated reduction, respectively. This indicates that acetoin can play a role in off-flavor formation in wines. Definitely, high concentrations of acetoin produced by non-*Saccharomyces* yeasts can be utilized by *S. cerevisiae* in mixed and sequential culture fermentations [54].

Non-*Saccharomyces* yeasts have also been reported to affect the concentration of polysaccharides in wine [55]. Polysaccharides can positively influence wine taste and mouth-feel by increasing the perception of wine "viscosity" and "fullness" on the palate [56]. The early death of some non-*Saccharomyces* yeasts during fermentation can also be a source of specific nutrients for *S. cerevisiae* enabling it to ferment optimally. These nutrients include cellular constituents such as cell wall polysaccharides (mannoproteins). For this method of nutrient supply to be effective, any killer or other inhibitory effects by the non-*Saccharomyces* yeasts against *S. cerevisiae* should be known [57] so that the subsequent *S. cerevisiae* fermentation is not adversely affected.

Other non-*Saccharomyces* extracellular enzymatic activities, such as proteolytic and pectinolytic (polygalacturonase) enzymes, might also be beneficial to winemaking [58]. For example, proteolytic activity of some non-*Saccharomyces* yeast could lead to a reduction in protein levels with accompanying increase in protein stability of the end-product. Species found to produce the greatest number of extracellular enzymes are *C. stellata*, *H. uvarum* and *M. pulcherrima* [59].

Certain flavor and aroma compounds are present in grapes as glycosidic precursors with no sensory properties [60]. These compounds may be hydrolyzed by the enzyme D-glucosidase to form free volatiles that can improve the flavor and aroma of wine, but this enzyme is not encoded by the *S. cerevisiae* genome [61]. However, certain non-*Saccharomyces* yeasts belonging to the genera *Debaryomyces*, *Hansenula*, *Candida*, *Pichia*, and *Hanseniaspora* possess various degrees of D-glucosidase activity [21] and can play a role in releasing volatile compounds from non-volatile precursors. An intracellular D-glucosidase has also been isolated and purified from *Debaryomyces hansenii*. This enzyme, which is not inhibited by glucose and ethanol, was used during fermentation of Muscat grape juice, resulting in an increase in concentration of monoterpenols in the wine [62].

4. Non-*Saccharomyces* Strains as Glycosidase Producers for Vinification

4.1. β-Glucosidases

Glycosidically-bound volatiles are highly complex and diverse, especially regarding the aglycone moiety. The sugar parts consist of β-D-glucopyranosides and different diglycosides: 6-*O*-α-L-arabinofuranosyl-β-D-glucopyranosides, 6-*O*-α-L-arabinopyranosyl-β-D-glucopyranosides (vicianosides), 6-*O*-α-L-rhamnopyranosyl-β-D-glucopyranoside (rutinosides), 6-*O*-β-D-apiofuranosyl-β-D-glucopyranosides, 6-*O*-β-D-glucopyranosyl-β-D-glucopyranosides, and 6-*O*-β-D-xilopyranosyl-β-D-glucopyranosides (primeverosides). The aglycon part is often formed with terpenols, but other flavor precursors can occur, such as linear or cyclic alcohols, C-13 norisoprenoids, phenolic acids, and probably volatile phenols such as vanillin [63].

If we consider only glycosides with the most flavorant aglycons the most abundant in grape juice are apiosylglycosides (up to 50% according to grape variety) followed by rutinosides (6% to 13%) and, finally, glucosides (4% to 9%). All glycosides are not present in all cultivars and their proportions also differ according to grapes [64]. The glycoside flavor potential from grapes remains quite stable during winemaking and in young wines as well. These findings opened a new field of intensive research on the chemistry of glycoconjugated flavor compounds to exploit this important flavor source. Some aglycones are already odorous when released from glycosides and can contribute to the varietal aroma of wines [65].

Terpene glycosides can be hydrolysed by an enzymic way [66] to enrich wine flavor by release of free aromatic compounds from natural glycoside precursors. Enzymatic hydrolysis of glycosides is carried out with various enzymes which act sequentially according to two steps: firstly, α-L-rhamnosidase, α-L-arabinosidase, or β-D-apiosidase make the cleavage of the terminal sugar and rhamnose, arabinose, or apiose and the corresponding β-D-glucosides are released; subsequent liberation of monoterpenol takes place after action of a β-D-glucosidase (Figure 1) [67]. Nevertheless, one-step hydrolysis of disaccharide glycosides has also been described; enzymes catalysing this reaction have been isolated from grapes [67]. Enzymic hydrolysis of glycoside extracts from Muscat, Riesling, Semillon, Chardonnay, Sauvignon, and Sirah varieties have provoked the liberation not only of terpenes, but also C-13 norisoprenoids, such as 3-oxo-β-ionol and 3-hydroxy-β-damascenona [68].

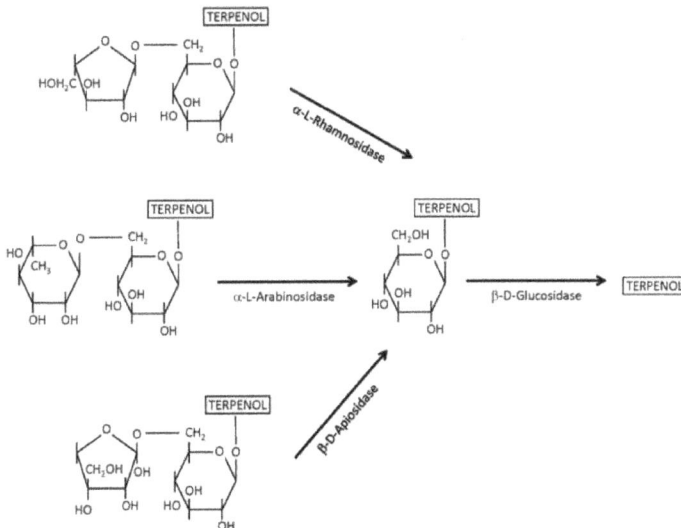

Figure 1. Sequential enzymatic hydrolysis of dissacharidic flavor precursors [66].

Few data are available regarding glycosidase activities of oenological yeast strains and the technological properties of the enzymes. Low α-rhamnosidase, α-arabinosidase, or β-apiosidase activities were detected in S. cerevisiae [69], but different efforts have been made to clone these genes obtained from differents microorganisms in S. cerevisiae [70]. Nevertheless, data on β-glucosidase activity on Saccharomyces are contradictory. First, results showed that these yeasts had a very low activity [71] but Delcroix et al. [69] found three enological strains showing high β-glucosidase activity. On the other hand, Darriet et al. [72] have shown that hydrolases located in the periplasmic space of a strain of S. cerevisiae were able to hydrolyse monoterpene glucosides of Muscat grapes; they also found that the activity of this β-glucosidase was glucose independent. Mateo and Di Stefano [73] detected β-glucosidase activity in different Saccharomyces strains on the basis of its hydrolytic activity on p-nitrophenyl-β-D-glucoside (pNPG) and terpene glucosides of Muscat juice. This enzymatic activity is induced by the presence of bound β-glucose as carbon source in the medium and seems to be a characteristic of the yeast strain. This β-glucosidase is associated with the yeast cell wall, is quite glucose independent but is inhibited by ethanol. This β-glucosidase is associated with the yeast cell wall is quite glucose independent but is inhibited by ethanol. These results could open new pathways regarding other glycosidase activities in S. cerevisiae; α-rhamnosidase, α-arabinosidase or β-apiosidase activities could be induced in wine yeast by changing the composition of the medium including inductive compounds, as well as in filamentous fungi [74].

Several flavor and aroma compounds in grapes are present as glycosylated flavorless precursors [2]. These compounds may be hydrolyzed by the enzyme β-glucosidase to form free volatiles that can increase the flavor and aroma of wine, but this enzyme is not encoded by the S. cerevisiae genome [46]. In contrast, non-Saccharomyces yeasts belonging to the genera Debaryomyces, Hansenula, Candida, Pichia, and Hanseniaspora possess various degrees of β-glucosidase activity and can play a role in releasing volatile compounds from non-volatile precursors [75]. Co-fermentation of Chardonnay grape juice with D. pseudopolymorphus and S. cerevisiae resulted in an increased concentration of the terpenols: citronellol, nerol, and geraniol in wine [76]. Similarly, cofermentation of Muscat grape juice with D. vanriji and S. cerevisiae produced wines with increased concentration of several terpenols [77]. Equally, mixed cultures of Sauvignon Blanc grape juice with C. zemplinina / S. cerevisiae and T. delbrueckii / S. cerevisiae produced wines with high concentrations of terpenols compared to wines only fermented with S. cerevisiae [23].

It has been reported that non-*Saccharomyces* yeasts can produce β-glucosidase [58]. The β-glucosidases from non-*Saccharomyces* species, such as C. molischiana, C. wickerhamii, and P. anomala were found to be more tolerant of winemaking conditions (for example, low pH values, low temperatures, high sugar, or ethanol levels) and tend to be more specific for glycosides than those from other yeast species [78]. Attempts have previously been made to enhance wine aroma using non-*Saccharomyces* yeasts and their glycosidases [79] because there is substantial yeast diversity in grapes and wines. Screening indigenous yeasts with glycosidases and their application in winemaking may allow wineries to make wines with more pleasant, typical, varietal aroma characteristics. Therefore, it is important to explore the potential of indigenous yeast biodiversity from specific enological ecosystems for specific and abundant β-glucosidases. β-Glucosidase-producing strains can be screened in Petri dishes with media containing cellulose-congo red, p-nitrophenyl-β-D-glucopyranoside (pNPG), or 4-methylumbelliferyl-β-dglucuronide (4-MUG) [80,81].

Yeasts of the *Hansenula* species isolated from fermenting must were reported to have an inducible β-glucosidase activity, but this enzyme was inhibited by glucose [82]. Other yeast strains, such as C. molischiana and C. wickerhamii, also possess activities towards various β-glucosides and they were little influenced by the nature of aglycon [83]. β-Glucosidase from C. molischiana was immobilized to Duolite A-568 resin, showing similar physicochemical properties to those of free enzyme. The immobilized enzyme was found to be very stable under wine conditions and could be used repeatedly for several hydrolyzes of bound aroma [84]. *Endomyces fibuliger* also produces extracellular β-glucosidase when grown in malt extract broth [85].

Screening 370 strains belonging to 20 species of yeasts, all of the strains of the species *D. castelli*, *D. hansenii*, *D. polymorphus*, *K. apiculata*, and *H. anomala* showed β-glucosidase activity [86]. A strain of *D. hansenii* exhibited the highest exocellular activity and some wall-bound and intracellular activity and its synthesis, occurred during exponential growth, was enhanced by aerobic conditions and repressed by high glucose concentration. The optimum condition for this enzyme was pH 4.0–5.0 and 40 °C. This enzyme was immobilized using a one-step procedure on hydroxyapatite. The immobilized enzyme exhibited a lower activity than the purified free enzyme, but was much more stable than the enzyme in cell-free supernatant [87]. Their studies have shown the ability of several wine yeasts to hydrolyse terpenoids, norisoprenoids and benzenoids glycosides; among wine yeasts *H. uvarum* was able to hydrolyze both glyco-conjugated forms of pyranic and furanic oxides of linalool [88]. Other authors have also shown the important role of non-*Saccharomyces* species in releasing the glycosidic-bound fraction of grape aroma components [89].

A total of 17 *Pichia* (*Wickerhamomyces*) isolates obtained from enological ecosystems in the Utiel-Requena Spanish region were characterized by physiological and molecular techniques (PCR-RFLP and sequencing) as belonging to the species *P. fermentans*, *P. membranifaciens*, and *W. anomalus*. In order to characterize their enzymatic abilities, xylanase, β-glucosidase, lipase, esterase, protease, and pectinase qualitative and quantitative assays were made. *W. anomalus* and *P. membranifaciens* showed to be the most interesting species to be used as sources of enzymes for the winemaking industry. Glycosidase enzymes show a high degree of tolerance to high levels of glucose and ethanol, making them of great interest to be used in enological procedure [90].

The sensorial characteristics of the wines produced with Muscat grapes are related to the level of terpene alcohols, so an improvement of such a level, as a result of hydrolytic processes conducted by *Hanseniaspora*, is expected. Isolates from *H. uvarum* and *H. vineae* have been proved to be good candidates to be used in commercial vinification processes to enhance wine properties. Wine inoculated with yeasts showed an increase in the level of aromatic compounds (Table 2) [91].

Table 2. Terpene compounds in Muscat wine. Concentration expressed as μg/L [a] [91].

Compound	Control [b]	*Hanseniaspora* Inoculated		
		H. uvarum H107	*H. vineae* G26	*H. vineae* P38
Oxide A [c]	29.7 (1.2)	30.4 (2.1)	33.7 (3.2)	26.9 (3.4)
Oxide B [d]	nd	nd	nd	nd
Linalool	20.0 (0.9)	40.4 * (3.9)	47.4 * (3.4)	38.2 * (5.3)
Ho-trienol	24.0 (3.2)	51.3 *(5.3)	35.1 * (4.2)	24.9 * (0.6)
2-Phenylethanol	1890.2 (43.4)	3057.5 * (39.8)	2747.8 * (26.8)	2568.5 * (45.6)
Oxide C [e]	nd	nd	nd	nd
Oxide D [f]	nd	nd	nd	nd
Terpineol	53.3 (3.4)	67.2 * (4.7)	65.1 *(1.2)	54.5 (3.9)
Nerol	24.6 (2.8)	25.8 (1.1)	23.4 (3.1)	26.3 (1.2)
Geraniol	59.8 (5.0)	61.3 (3.7)	56.9 (1.7)	62.8 (1.7)
Diol 1 [g]	43.2 (4.7)	87.9 * (2.1)	80.2 * (2.1)	81.2 * (3.2)
4-Vinylphenol	63.2 (1.2)	89.7 * (2.4)	75.7 * (5.8)	62.1 (0.9)
Endiol [h]	nd	58.8 * (2.1)	52.0 * (3.4)	34.1 * (4.2)
Diol 2 [i]	12.0 (0.6)	13.4 (0.9)	7.8 (2.6)	10.1 (0.9)
2-Phenylethyl acetate	28.0 (4.1)	56.2 * (7.2)	23.3 (1.2)	25.8 (4.7)
2-Methoxy-4-vinylphenol	89.0 (6.1)	103.0 * (5.3)	105.4 * (6.5)	94.1 (2.9)

[a] Values in brackets represent standard deviation ($n = 3$). ANOVA one factor, significant difference is indicated as * ($p < 0.05$); [b] Wine produced only with *Saccharomyces cerevisiae*; [c] *cis*-5-vinyltetrahydro-1, 1,5-trimethyl-2-furanmethanol; [d] *trans*-5-vinyltetrahydro-1,1,5-trimethyl-2-furanmethanol; [e] *cis*-6-vinyltetrahydro-2,2,6-trimethyl-2H-pyran-3-ol; [f] *trans*-6-vinyltetrahydro-2,2,6-trimethyl-2H-pyran-3-ol; [g] 2,6-Dimethyl-3,7-octadien-2,6-diol; [h] 2,6-Dimethyl-7-octene-2,6-diol; [i] 2,6-Dimethyl-2,7-octadien-1,6-diol; nd: not detected

The potential applications of wild yeast strains with β-glucosidase activity have been investigated under simulated oenological conditions, coupled with the exploration of the potential applications of the β-glucosidases by studying the enzymatic activity and stability under similar oenological conditions [92]. The effects of different oenological factors on β-glucosidase production indicated that one isolate from the *T. asahii* strain had higher β-glucosidase production than the other strains under low pH conditions. However, isolates from *H. uvarum* and *S. cerevisiae* strain showed higher β-glucosidase production under high-sugar conditions. Furthermore, the influence of oenological factors on the activity and stability of the β-glucosidases revealed that the enzyme from the *T. asahii* strain had a stronger low-pH-value resistance than the other yeast β-glucosidases [92].

Hu et al. [93] applied a semiquantitative colorimetric assay to screen yeasts from three different regions of China. Among 493 non-*Saccharomyces* isolates belonging to eight genera, three isolates were selected for their high levels of β-glucosidase activity and were identified as *H. uvarum*, *P. membranifaciens*, and *Rhodotorula mucilaginosa*. β-Glucosidase from the *H. uvarum* strain showed the highest activity in winemaking conditions among the selected isolates. For aroma enhancement in winemaking, the glycosidase extract from *H. uvarum* exhibited catalytic specificity for aromatic glycosides of C13-norisoprenoids and some terpenes, enhancing fresh floral, sweet, berry, and nutty aroma characteristics in wine.

4.2. Xylanases

β-1,4-xylan is a heteroglycan with a backbone of β-(1→4)-linked D-xylopyranose residues that can be substituted with L-arabinofuranose, D-glucuronic acid, and/or 4-O-methyl-D-glucuronic acid [94]. It constitutes the major component of hemicelluloses found in the cell walls of monocots and hardwoods, and represents one of the most abundant biomass resources. Recently, xylanolytic enzymes of microbial origin have received great attention due to their possible industrial applications for sustainable fuel-ethanol production from xylan. Two key reactions proceed during hydrolysis of the xylan backbone; endo-1,4-β-xylanases (1,4-β-D-xylan xylanohydrolase) hydrolyze internal β-(1→4)-xylosidic linkages in the insoluble xylan backbone to yield soluble xylooligosaccharides, while 1,4-β-xylosidases are exoglycosidases that cleave terminal xylose monomers from the non-reducing end of short-chain xylooligosaccharides [95]. Additional enzyme activities, such as α-L-arabinofuranosidase, α-D-glucuronidase, and acetyl xylan esterase, remove side-chain substituents. 1,4-β-Xylosidase is important in xylan degradation, considering that xylans are not completely hydrolyzed by xylanases alone. The finding of new isolates of non-*Saccharomyces* yeasts, showing beneficial enzymes (such as β-glucosidase and β-xylosidase) can contribute to the production of quality wines [96]. In a selection and characterization program we have studied 114 isolates of non-*Saccharomyces* yeasts, four isolates were selected because of their both high β-glucosidase and β-xylosidase activities. The ribosomal D1/D2 regions were sequenced to identify them as *P. membranifaciens*, *H. vineae*, *H. uvarum*, and *W. anomalus*. The induction process was optimized to be carried on YNB-medium supplemented with 4% xylan, inoculated with 10^6 cfu/mL and incubated 48 h at 28 °C without agitation. Most of the strains had a pH optimum of 5.0 to 6.0 for both the β-glucosidase and β-xylosidase activities. The effect of sugars was different for each isolate and activity. Each isolate showed a characteristic set of inhibition, enhancement or null effect for β-glucosidase and β-xylosidase. The volatile compounds liberated from wine incubated with each of the four yeasts were also studied, showing an overall terpene increase when wines were treated with non-*Saccharomyces* isolates. In detail, terpineol, 4-vinyl-phenol, and 2-methoxy-4-vinylphenol increased after the addition of *Hanseniaspora* isolates. Wines treated with *Hanseniaspora*, *Wickerhamomyces*, or *Pichia* produced more 2-phenyl ethanol than those inoculated with other yeasts [97].

An ethanol-tolerant 1,4-β-xylosidase was purified from cultures of a strain of *P. membranifaciens* grown on xylan at 28 °C. The enzyme was purified by sequential chromatography on DEAE-cellulose and Sephadex G-100. The relative molecular mass of the enzyme was determined to be 50 kDa by SDS-PAGE. The activity of 1,4-β-xylosidase was optimum at pH 6.0 and 35 °C. The activity had a Km of 0.48 ± 0.06 mmol·L^{-1} and a Vmax of 7.4 ± 0.1 μmol·min^{-1}·mg^{-1} protein for p-nitrophenyl-β-D-xylopyranoside. The enzyme characteristics (pH and thermal stability, low inhibition rate by glucose and ethanol tolerance) make this enzyme a good candidate to be used

Fermentation **2016**, *2*, 14

in enzymatic production of xylose and improvement of hemicellulose saccharification for production of bioethanol [98].

5. Conclusions

It is generally accepted that the wealth of yeast biodiversity with hidden potential, especially for oenology, is largely untapped. In order to exploit the potential benefits of non-*Saccharomyces* yeasts in wine production and to minimize potential spoilage, the yeast populations on grapes and in must, as well as the effect of wine-making practices on these yeasts, must be known, as must the metabolic characteristics of non-*Saccharomyces* yeasts. Strain selection will be very important, as not all strains within a species will necessarily show the same desirable characteristics. For example, significant variability is found in the formation of β-glucosidase amongst strains within some non-*Saccharomyces* yeast species.

Whatever the outcome of the search for non-*Saccharomyces* yeasts for use in wine production, the accepted list of desirable characteristics as pertaining to the wine yeast *S. cerevisiae* will not necessarily apply to non-*Saccharomyces* yeasts. High fermentation efficiency, sulfite tolerance and killer properties, for example, might not be needed in the new technology of wine production. The new non-*Saccharomyces* wine yeasts will necessarily have a different list of desired characteristics: efficient sugar utilization, enhanced production of desirable volatile esters, enhanced liberation of grape terpenoids and production of glycerol to improve wine flavor and other sensory properties can be met by selected non-*Saccharomyces* wine yeasts.

Acknowledgments: The Universitat de València kindly supported this work included in the project "Identification and biotechnological characterization of yeasts isolated from agrifood residues of the Valencian Community", grant INV-AE-336499.

Author Contributions: Sergi Maicas and José Juan Mateo wrote the paper.

Conflicts of Interest: The authors declare no conflict of interest.

References

1. Padilla, B.; Gil, J.V.; Manzanares, P. Past and future of non-*Saccharomyces* yeasts: From spoilage microorganisms to biotechnological tools for improving wine aroma complexity. *Front. Microbiol.* **2016**. [CrossRef] [PubMed]
2. Pretorius, I.S. Tailoring wine yeast for the new millennium: Novel Approaches to the ancient art of winemaking. *Yeast* **2000**, *16*, 675–729. [CrossRef]
3. Jackson, R.S. *Wine Science: Principles and Applications*, 1st ed.; Elsevier: San Diego, CA, USA, 1994.
4. Andorrà, I.; Berradre, M.; Rozès, N.; Mas, A.; Guillamón, J.M.; Esteve-Zarzoso, B. Effect of pure and mixed Cultures of the Main Wine yeast Species on Grape Must Fermentations. *Eur. Food Res. Technol.* **2010**, *231*, 215–224. [CrossRef]
5. Gil, J.V.; Mateo, J.J.; Jiménez, M.; Pastor, A.; Huerta, T. Aroma compounds in wine as influenced by apiculate yeasts. *J. Food Sci.* **1996**, *61*, 1247–1250. [CrossRef]
6. Wang, C.; Mas, A.; Esteve, B. The interaction between *Saccharomyces cerevisiae* and non-*Saccharomyces* yeast during alcoholic fermentation is species and strain specific. *Front. Microbiol.* **2016**. [CrossRef] [PubMed]
7. Mateo, J.J.; Jiménez, M.; Huerta, T.; Pastor, A. Contribution of different yeasts isolated from musts of Monastrell grapes to the aroma of wine. *Int. J. Food Microbiol.* **1991**, *14*, 153–160. [CrossRef]
8. Rodríguez-Gómez, F.; Arroyo-López, F.N.; López-López, A.; Bautista-Gallego, J.; Garrido-Fernández, A. Lipolytic activity of the yeast species associated with the fermentation/storage phase of ripe olive processing. *Food Microbiol.* **2010**, *27*, 604–612. [CrossRef] [PubMed]
9. Heard, G. Novel yeasts in winemaking—Looking to the future. *Food Aust.* **1999**, *51*, 347–352.
10. Barata, A.; Malfeito-Ferreira, M.; Loureiro, V. The microbial ecology of wine grape berries. *Int. J. Food Microbiol.* **2012**, *153*, 243–259. [CrossRef] [PubMed]
11. Martini, A. Origin and domestication of the wine yeast *Saccharomyces cerevisiae*. *J. Wine Res.* **1993**, *4*, 165–176. [CrossRef]

12. Ribéreau-Gayon, P.; Dubordieu, D.; Donèche, B.; Lonvaud, A. *Handbook of Enology Volume 1: The Microbiology of Wine and Vinifications*; John Wiley & Sons Ltd.: Chichester, UK, 2000.

13. Johnson, E.A. Biotechnology of non-*Saccharomyces* yeasts—The Ascomycetes. *Appl. Microbiol. Biotechnol.* **2013**, *97*, 503–517. [CrossRef] [PubMed]

14. Pretorius, I.S.; Van Der Westhuizen, T.J.; Augustyn, O.P.H. Yeast biodiversity in vineyards and wineries and its importance to the South African wine industry—A review. *S. Afr. J. Enol. Vitic.* **1999**, *20*, 61–75.

15. Kurtzman, C.P. *The Yeasts: A Taxonomic Study*, 5th ed.; Elsevier Science: Waltham, MA, USA, 2012.

16. Davenport, R.R. Microecology of yeast and yeastlike organisms associated with and English vineyard. *Vitis* **1974**, *13*, 123–130.

17. Jolly, N.P.; Vera, C.; Pretorius, I.S. Not your ordinary yeast: Non-Saccharomyces yeasts in wine production uncovered. *FEMS Yeast Res.* **2014**, *14*, 215–237. [CrossRef] [PubMed]

18. Boynton, P.J.; Duncan, G. The ecology and evolution of non-domesticated *Saccharomyces* species. *Yeast* **2014**, *12*, 449–462.

19. Henick-Kling, T.; Edinger, W.; Daniel, P.; Monk, P. Selective effects of sulfur dioxide and yeast starter culture addition on indigenous yeast populations and sensory characteristics of the wine. *J. Appl. Microbiol.* **1998**, *84*, 865–876. [CrossRef]

20. Jolly, N.P.; Augustyn, O.H.P.; Pretorius, I.S. The effect of non-*Saccharomyces* yeasts on fermentation and wine quality. *S. Afr. J. Enol. Vitic.* **2003**, *24*, 55–62.

21. Romano, P.; Suzzi, G.; Domizio, P.; Fatichenti, F. Secondary products formation as a tool for discriminating non-*Saccharomyces* wine strains. *Antonie Leeuwenhoek* **1997**, *71*, 239–242. [CrossRef] [PubMed]

22. Ciani, M.; Comitini, F. Non-*Saccharomyces* wine yeasts have a promising role in biotechnological approaches to winemaking. *Ann. Microbiol.* **2011**, *61*, 25–32. [CrossRef]

23. Sadoudi, M.; Tourdot-Maréchal, R.; Rousseaux, S.; Steyer, D.; Gallardo-Chacón, J.J.; Ballester, J.; Vichi, S.; Guérin-Schneider, R.; Caixach, J.; Alexandre, H. Yeast-yeast interactions revealed by aromatic profile analysis of Sauvignon Blanc wine fermented by single or co-culture of non-*Saccharomyces* and *Saccharomyces* yeasts. *Food Microbiol.* **2012**, *32*, 243–253. [CrossRef] [PubMed]

24. Sun, L.; Hebert, A.S.; Yan, X.; Zhao, Y.; Westphall, M.S.; Rush, M.J.P.; Zhu, G.; Champion, M.M.; Coon, J.J.; Dovichi, N.J. Over 10000 peptide identifications from the HeLa proteome by using single-shot capillary zone electrophoresis combined with tandem mass spectrometry. *Angew. Chem.* **2014**, *126*, 14151–14153. [CrossRef]

25. Comitini, F.; Gobbi, M.; Domizio, P.; Romani, C.; Lencioni, L.; Mannazzu, I.; Ciani, M. Selected non-*Saccharomyces* wine yeasts in controlled multistarter fermentations with *Saccharomyces cerevisiae*. *Food Microbiol.* **2011**, *28*, 873–882. [CrossRef] [PubMed]

26. Maturano, Y.P.; Rodríguez Assaf, L.A.; Toro, M.E.; Nally, M.C.; Vallejo, M.; Castellanos de Figueroa, L.I.; Combina, M.; Vazquez, F. Multi-enzyme production by pure and mixed cultures of *Saccharomyces* and non-*Saccharomyces* yeasts during wine fermentation. *Int. J. Food Microbiol.* **2012**, *155*, 43–50. [CrossRef] [PubMed]

27. Kutyna, D.; Varela, C.; Henschke, P.; Chambers, P.; Stanley, G. Microbiological approaches to lowering ethanol concentration in wine. *Trends Food Sci. Technol.* **2010**, *21*, 293–302. [CrossRef]

28. Di Maio, S.; Polizzotto, G.; Di Gangi, E.; Foresta, G.; Genna, G.; Verzera, A. Biodiversity of indigenous *Saccharomyces* populations from old wineries of South-Eastern Sicily (Italy): Preservation and economic potential. *PLoS ONE* **2012**. [CrossRef]

29. Gonzalez, R.; Quirós, M.; Morales, P. Yeast respiration of sugars by non-*Saccharomyces* yeast species: A promising and barely explored approach to lowering alcohol content of wines. *Trends Food Sci. Technol.* **2013**, *29*, 55–61. [CrossRef]

30. Erten, H.; Tanguler, H. Influence of *Williopsis saturnus* yeasts in combination with *Saccharomyces cerevisiae* on wine fermentation. *Lett. Appl. Microbiol.* **2010**, *50*, 474–479. [CrossRef] [PubMed]

31. Tilloy, V.; Ortiz-Julien, A.; Dequin, S. Reduction of ethanol yield and improvement of glycerol formation by adaptive evolution of the wine yeast *Saccharomyces* cerevisiae under hyperosmotic conditions. *Appl. Environ. Microbiol.* **2014**, *80*, 2623–2632. [CrossRef] [PubMed]

32. Ciani, M.; Comitini, F.; Mannazzu, I.; Domizio, P. Controlled mixed culture fermentation: A new perspective on the use of non-*Saccharomyces* yeasts in winemaking. *FEMS Yeast Res.* **2010**, *10*, 123–133. [CrossRef] [PubMed]

33. Quiros, M.; Rojas, V.; Gonzalez, R.; Morales, P. Selection of non-*Saccharomyces* yeast strains for reducing alcohol levels in wine by sugar respiration. *Int. J. Food Microbiol.* **2014**, *181*, 85–91. [CrossRef] [PubMed]

34. Gobbi, M.; De Vero, L.; Solieri, L.; Comitini, F.; Oro, L.; Giudici, P.; Ciani, M. Fermentative aptitude of non-*Saccharomyces* wine yeast for reduction in the ethanol content in wine. *Eur. Food Res. Technol.* **2014**, *239*, 41–48. [CrossRef]

35. Fleet, G.H. Wine yeasts for the future. *FEMS Yeast Res.* **2008**, *8*, 979–995. [CrossRef] [PubMed]

36. Styger, G.; Prior, B.; Bauer, F.F. Wine flavor and aroma. *J. Ind. Microbiol. Biotechnol.* **2011**, *38*, 1145–1159. [CrossRef] [PubMed]

37. Swiegers, J.H.; Bartowsky, E.J.; Henschke, P.A.; Pretorius, I.S. Yeast and bacterial modulation of wine aroma and flavour. *Aust. J. Grape Wine Res.* **2005**, *11*, 139–173. [CrossRef]

38. Zironi, R.; Romano, P.; Suzzi, G.; Battistutta, F.; Comi, G. Volatile metabolites produced in wine by mixed and sequentialcultures of *Hanseniaspora guilliermondii* or *Kloeckera apiculata* and *Saccharomyces cerevisiae*. *Biotechnol. Lett.* **1993**, *15*, 235–238. [CrossRef]

39. Arévalo-Villena, M.; Úbeda-Iranzo, J.F.; Gundllapalli, S.B.; Cordero-Otero, R.R.; Briones-Pérez, A.I. Characterization of an exocellular β-glucosidase from *Debaryomyces pseudopolymorphus*. *Enzyme Microb. Technol.* **2006**, *39*, 229–234. [CrossRef]

40. Ocón, E.; Gutiérrez, A.R.; Garijo, P.; López, R.; Santamaría, P. Presence of non-*Saccharomyces* yeasts in cellar equipment and grape juice during harvest time. *Food Microbiol.* **2010**, *27*, 1023–1027. [CrossRef] [PubMed]

41. Clemente-Jiménez, J.M.; Mingorance-Cazorla, L.; Martínez-Rodríguez, S.; Las Heras-Vázquez, F.J.; Rodríguez-Vico, F. Molecular characterization and oenological properties of wine yeasts isolated during spontaneous fermentation of six varieties of grape must. *Food Microbiol.* **2004**, *21*, 149–155. [CrossRef]

42. Arévalo-Villena, M.; Úbeda-Iranzo, J.F.; Briones-Pérez, A.I. β-Glucosidase activity in wine yeasts: Application in enology. *Enzyme Microb. Technol.* **2007**, *40*, 420–425. [CrossRef]

43. Jackson, R.S. *Wine Science: Principles, Practice, Perception*, 2nd ed.; Elsevier: San Diego, CA, USA, 2000.

44. Lambrechts, M.G.; Pretorius, I.S. Yeast and its importance to wine aroma–A review. *S. Afr. J. Enol. Vitic.* **2000**, *21*, 97–129.

45. Swiegers, J.H.; Pretorius, I.S. Yeast modulation of wine flavor. *Adv. Appl. Microbiol.* **2005**, *57*, 131–175. [PubMed]

46. Ubeda-Iranzo, J.F.; Briones-Perez, A.I.; Izquierdo-Cañas, P.M. Study of the oenological characteristics and enzymatic activities of wine yeasts. *Food Microbiol.* **1998**, *15*, 399–406. [CrossRef]

47. Scanes, K.T.; Hohmann, S.; Prior, B.A. Glycerol production by the yeast *Saccharomyces* cerevisiae and its relevance to wine: A review. *S. Afr. J. Enol. Vitic.* **1998**, *19*, 17–24.

48. Ciani, M.; Maccarelli, F. Oenological properties of non-*Saccharomyces* yeasts associated with wine-making. *World J. Microbiol. Biotechnol.* **1998**, *14*, 199–203. [CrossRef]

49. Prior, B.A.; Toh, T.H.; Jolly, N.; Baccari, C.L.; Mortimer, R.K. Impact of yeast breeding for elevated glycerol production on fermentation activity and metabolite formation in Chardonnay. *S. Afr. J. Enol. Vitic.* **2000**, *21*, 92–99.

50. Swiegers, J.H.; Kievit, R.L.; Siebert, T.; Lattey, K.A.; Bramley, B.R.; Francis, I.L. The influence of yeast on the aroma of Sauvignon Blanc wine. *Food Microbiol.* **2009**, *26*, 204–211. [CrossRef] [PubMed]

51. Anfang, N.; Brajkovich, M.; Goddard, M.R. Co-fermentation with *Pichia kluyveri* increases varietal thiol concentrations in Sauvignon Blanc. *Aust. J. Grape Wine Res.* **2009**, *15*, 1–8. [CrossRef]

52. Zott, K.; Thibon, C.; Bely, M.; Lonvaud-Funel, A.; Dubourdieu, D.; Masneuf-Pomarede, I. The grape must non-*Saccharomyces* microbial community: Impact on volatile thiol release. *Int. J. Food Microbiol.* **2011**, *151*, 210–215. [CrossRef] [PubMed]

53. Romano, P.; Suzzi, G. Higher alcohol and acetoin production by *Zygosaccharomyces* wine yeasts. *J. Appl. Bacteriol.* **1993**, *75*, 541–545. [CrossRef]

54. Mateo, J.J.; Peris, L.; Ibañez, C.; Maicas, S. Characterization of glycolytic activities from non-*Saccharomyces* yeasts isolated from Bobal musts. *J. Ind. Microbiol. Biotechnol.* **2011**, *38*, 347–354. [CrossRef] [PubMed]

55. Domizio, P.; Liu, Y.; Bisson, L.F.; Barile, D. Use of non-*Saccharomyces* wine yeasts as novel sources of mannoproteins in wine. *Food Microbiol.* **2014**, *43*, 5–15. [CrossRef] [PubMed]

56. Vidal, S.; Francis, L.; Noble, A.; Kwiatkowski, M.; Cheynier, V.; Waters, E. Taste and mouth-feel properties of different types of tannin-like polyphenolic compounds and anthocyanins in wine. *Anal. Chim. Acta* **2004**, *513*, 57–65. [CrossRef]

57. Fleet, G.H.; Prakitchaiwattana, C.; Beh, A.L.; Heard, G. The yeast ecology of wine grapes. In *Biodiversity and Biotechnology of Wine Yeasts*; Ciani, M., Ed.; Research Signpost: Kerala, India, 2002; pp. 1–17.

58. Strauss, M.L.; Jolly, N.P.; Lambrechts, M.G.; Rensburg, P.V. Screening for the production of extracellular hydrolytic enzymes by non-*Saccharomyces* wine yeasts. *J. Appl. Microbiol.* **2001**, *91*, 182–190. [CrossRef] [PubMed]

59. Mateo, J.J.; Maicas, S.; Thießen, C. Biotechnological characterisation of exocellular proteases produced by enological *Hanseniaspora* isolates. *Int. J. Food Sci. Technol.* **2015**, *50*, 218–225. [CrossRef]

60. Pretorius, I.S. The genetic analysis and tailoring of wine yeasts. In *Functional Genetics of Industrial Yeasts*; De Winde, J.H., Ed.; Springer-Verlag: Berlin, Germany, 2003; Volume 2, pp. 99–142.

61. Van Rensburg, P.; Stidwell, T.; Lambrechts, M.G.; Cordero-Otero, R.R.; Pretorius, I.S. Development and assessment of a recombinant *Saccharomyces cerevisiae* wine yeast producing two aroma-enhancing β-glucosidases encoded by the *Saccharomycopsis fibuligera* BGL_1 and BGL_2 genes. *Anal. Microbiol.* **2005**, *55*, 33–42.

62. Yanai, T.; Sato, M. Isolation and properties of β-glucosidase produced by *Debaryomyces hansenii* and its application in winemakig. *Am. J. Enol. Vitic.* **1999**, *50*, 231–235.

63. Sarry, J.E.; Gunata, Z. Plant and microbial glycoside hydrolases: Volatile release from glycosidic aroma precursors. *Food Chem.* **2004**, *87*, 509–521. [CrossRef]

64. Bayonove, C.; Gunata, Y.; Sapis, J.C.; Baumes, R.L.; Dugelay, I.; Grassin, C. L'aumento degli aromi nel vino mediante l'uso degli enzimi. *Vignevini* **1993**, *9*, 33–36. (In Italian)

65. Mateo, J.J.; Jiménez, M. Monoterpenes in grape juice and wines. *J. Chromatogr. A* **2000**, *881*, 557–567. [CrossRef]

66. Maicas, S.; Mateo, J.J. Hydrolysis of terpenyl glycosides in grape juice and other fruit juices: A review. *Appl. Microbiol. Biotechnol.* **2005**, *67*, 322–335. [CrossRef] [PubMed]

67. Gunata, Z.; Bitteur, S.; Brillouet, J.M.; Bayonove, C.; Cordonnier, R. Sequential enzymatic hydrolysis of potentially aromatic glycosides form grapes. *Carbohydr. Res.* **1988**, *184*, 139–149. [CrossRef]

68. Gunata, Z.; Bayonove, C.; Tapiero, C.; Cordonnier, R. Hydrolysis of grape monoterpenyl β-D-glucosides by various β-glucosidases. *J. Agric. Food Chem.* **1990**, *38*, 1232–1236. [CrossRef]

69. Delcroix, A.; Gunata, Z.; Sapis, J.C.; Salmon, J.M.; Bayonove, C. Glycosidase activities of three enological yeast strains during wine making. Effect on the terpenol content of Muscat wine. *Am. J. Enol. Vitic.* **1994**, *45*, 291–296.

70. Bisson, L.F.; Karpel, J.E. Genetics of yeast impacting wine quality. *Ann. Rev. Food Sci. Technol.* **2010**, *1*, 139–162. [CrossRef] [PubMed]

71. Gunata, Z.; Dugelay, I.; Sapis, J.C.; Baumes, R.; Bayonove, C. Action des glycosidases exogènes au cours de la vinification: Liberation de l'arôme à partir des précurseurs glycosidiques. *J. Int. Sci. Vigne Vin* **1990**, *24*, 133–144. (In French)

72. Darriet, P.; Boidron, J.N.; Dubourdieu, D. L'hydrolyse des hétérosides terpéniques du Muscat a Petit Grains par les enzymes périplasmiques de *Saccharomyces cerevisiae*. *Connaiss. Vigne Vin* **1988**, *22*, 189–195. (In French)

73. Mateo, J.J.; Di Stefano, R. Enological properties of β-glucosidase in wine yeasts. *Food Microbiol.* **1998**, *14*, 583–591. [CrossRef]

74. Dupin, I.; Gunata, Z.; Sapis, J.C.; Bayonove, C.; M'Bairaroua, O.; Tapiero, C. Production of a β-apiosidase by *Aspergillus niger*. Partial purification, properties and effect on terpenyl apiosylglucosides from grape. *J. Agric. Food Chem.* **1992**, *40*, 1886–1891. [CrossRef]

75. Spagna, G.; Barbagallo, R.N.; Palmeri, R.; Restuccia, C.; Giudici, P. Properties of endogenous β-glucosidase of a *Saccharomyces cerevisiae* strain isolated from Sicilian musts and wines. *Enzyme Microb. Technol.* **2002**, *31*, 1030–1035. [CrossRef]

76. Cordero Otero, R.R.; Ubeda Iranzo, J.F.; Briones-Perez, A.I.; Potgieter, N.; Villena, M.A.; Pretorius, I.S.; van Rensburg, P. Characterization of the β-glucosidase activity produced by enological strains of non-*Saccharomyces* yeasts. *J. Food Sci.* **2003**, *68*, 2564–2569. [CrossRef]

77. García, A.; Carcel, C.; Dulau, L.; Samson, A.; Aguera, E.; Agosin, E. Influence of a mixed culture with *Debaryomyces vanriji* and *Saccharomyces cerevisiae* on the volatiles of a Muscat wine. *J. Food Sci.* **2002**, *67*, 1138–1143. [CrossRef]

78. Jutaporn, S.; Sukanda, V.; Christian, E.B.; Kanit, V. The characterisation of a novel *Pichia anomala* β-glucosidase with potentially aroma-enhancing capabilities in wine. *Ann. Microbiol.* **2009**, *59*, 335–343.

79. Carrascosa, A.V.; Martinez-Rodriguez, A.; Cebollero, E.; Gonzalez, R. Molecular Wine Microbiology. In *Saccharomyces Yeast II: Secondary Fermentation*; Carrascosa, A.V., Muñoz, R., Gonzalez, R., Eds.; Elsevier: London, UK, 2011; pp. 33–49.
80. Hernandez, L.F.; Espinosa, J.C.; Fernandez, G.M.; Briones, A. β-glucosidase activity in a *Saccharomyces cerevisiae* wine strain. *Int. J. Food Microbiol.* **2003**, *80*, 171–176. [CrossRef]
81. Wang, Y.X.; Zhang, C.; Li, J.M.; Xu, Y. Different influences of β-glucosidases on volatile compounds and anthocyanins of Cabernet Gernischt and possible reason. *Food Chem.* **2013**, *140*, 245–254. [CrossRef] [PubMed]
82. Grossmann, M.; Rapp, A.; Rieth, W. Enzymatische Freisetzung gebundener Aromastoffe in Wein. *Dtsch. Lebensm. Rundsch.* **1987**, *83*, 7–12. (In German)
83. Gunata, Z.; Brillouet, J.M.; Voirin, S.; Baumes, B.; Cordonnier, R. Purification and some properties of an α-L-arabinofuranosidase from *Aspergillus niger*. Action on grape monoterpenyl arabinofuranosyl glucosidases. *J. Agric. Food Chem.* **1990**, *38*, 772–776. [CrossRef]
84. Gueguen, Y.; Chemardin, P.; Pien, S.; Arnaud, A.; Galzy, P. Enhancement of aromatic quality of Muscat wine by the use of immobilized β-glucosidase. *J. Biotechnol.* **1997**, *55*, 151–156. [CrossRef]
85. Brimer, L.; Nout, M.J.R.; Tuncel, G. Glycosidase (amygdalase and linamarase) from *Endomyces fibuliger* (LU677): Formation and crude enzyme properties. *Appl. Microbiol. Biotechnol.* **1998**, *49*, 182–188. [CrossRef] [PubMed]
86. Rosi, I.; Vinella, M.; Domizio, P. Characterization of β-glucosidase activity in yeasts of oenological origin. *J. Appl. Bacteriol.* **1994**, *77*, 519–527. [CrossRef] [PubMed]
87. Riccio, P.; Rossano, R.; Vinella, M.; Domizio, P.; Zito, F.; Sanseverino, F.; D'Elia, A.; Rosi, I. Extraction and immobilization in one step of two β-glucosidases released from a yeast strain of *Debaryomyces hansenii*. *Enzyme Microb. Technol.* **1999**, *24*, 123–129. [CrossRef]
88. Fernández-González, M.; Di Stefano, R.; Briones, A.I. Hydrolysis and transformation of terpene glycosides from Muscat must by different yeast species. *Food Microbiol.* **2003**, *20*, 35–41. [CrossRef]
89. Mendes, A.; Climaco, M.C.; Mendes, A. The role of non-*Saccharomyces* species in releasing glycosidic bound fraction of grape aroma components–A preliminary study. *J. Appl. Microbiol.* **2001**, *91*, 67–71. [CrossRef]
90. Madrigal, T.; Maicas, S.; Mateo, J.J. Glucose and ethanol tolerant enzymes produced by *Pichia* (*Wickerhamomyces*) isolates from enological ecosystems. *Am. J. Enol. Vitic.* **2013**, *64*, 126–133. [CrossRef]
91. Lopez, S.; Mateo, J.J.; Maicas, S. Characterization of *Hanseniaspora* isolates with potential aroma enhancing properties in Muscat wines. *S. Afr. J. Enol. Vitic.* **2014**, *35*, 292–303.
92. Wang, Y.; Xu, Y.; Li, J. A Novel extracellular β-glucosidase from *Trichosporon asahii*: Yield prediction, evaluation and application for aroma enhancement of cabernet sauvignon. *J. Food Sci.* **2012**, *77*, 505–515. [CrossRef] [PubMed]
93. Hu, K.; Qin, Y.; Tao, Y.S.; Zhu, X.L.; Peng, C.T.; Ullah, N. Potential of glycosidase from non-*Saccharomyces* isolates for enhancement of wine aroma. *J. Food Sci.* **2016**, *81*, 935–943. [CrossRef] [PubMed]
94. Bhat, M.K.; Hazlewood, G.P. Enzymology and other characteristics of cellulases and xylanases. In *Enzymes in Farm Animal Nutrition*; Bedford, M., Partridge, G., Eds.; CABI Publishing: Oxon, UK, 2001.
95. Polizeli, M.L.; Rizzati, A.C.; Monti, R.H.F.; Terenzi, C.G.; Jorge, J.A.; Amorin, D.S. Xylanases from fungi: Properties and industrial applications. *Appl. Microbiol. Biotechnol.* **2005**, *67*, 577–591. [CrossRef] [PubMed]
96. Linden, T.; Hahn-Hagerdal, B. Fermentation of lignocellulose hydrolysates with yeasts and xylose isomerase. *Enzyme Microb. Technol.* **1989**, *11*, 583589. [CrossRef]
97. López, M.C.; Mateo, J.J.; Maicas, S. Screening of β-glucosidase and β-xylosidase activities in four non-*Saccharomyces* yeast isolates. *J. Food Sci.* **2015**, *80*, 1696–1704. [CrossRef] [PubMed]
98. Romero, A.M.; Mateo, J.J.; Maicas, S. Characterization of an ethanol-tolerant 1,4-β-xylosidase produced by *Pichia membranifaciens*. *Lett. Appl. Microbiol.* **2012**, *55*, 354–361. [CrossRef] [PubMed]

fermentation

MDPI

Article

Batch Fermentation Options for High Titer Bioethanol Production from a SPORL Pretreated Douglas-Fir Forest Residue without Detoxification

Mingyan Yang [1,2,3], Hairui Ji [2,4] and J.Y. Zhu [2,*]

[1] School Environment Science Engineering, Chang'an University, Xi'an 710064, China; yangmingyan67@163.com
[2] USDA Forest Service, Forest Products Laboratory, Madison, WI 53726, USA; jzhu@fs.fed.us
[3] Key Laboratory of Subsurface Hydrology and Ecological Effects in Arid Region, Ministry of Education, Chang'an University, Xi'an 710064, China
[4] College of Life Science and Technology, Beijing University of Chemical Technology, Beijing 100029, China
[*] Correspondence: jzhu@fs.fed.us; Tel.: +1-608-231-9520

Academic Editor: Ronnie G. Willaert
Received: 3 June 2016; Accepted: 1 August 2016; Published: 11 August 2016

Abstract: This study evaluated batch fermentation modes, namely, separate hydrolysis and fermentation (SHF), quasi-simultaneous saccharification and fermentation (Q-SSF), and simultaneous saccharification and fermentation (SSF), and fermentation conditions, i.e., enzyme and yeast loadings, nutrient supplementation and sterilization, on high titer bioethanol production from SPORL-pretreated Douglas-fir forest residue without detoxification. The results indicated that Q-SSF and SSF were obviously superior to SHF operation in terms of ethanol yield. Enzyme loading had a strong positive correlation with ethanol yield in the range studied. Nutrient supplementation and sterility were not necessary for ethanol production from SPORL-pretreated Douglas-fir. Yeast loading had no substantial influence on ethanol yield for typical SSF conditions. After 96 h fermentation at 38 °C on shake flask at 150 rpm, terminal ethanol titer of 43.2 g/L, or 75.1% theoretical based on untreated feedstock glucan, mannan, and xylan content was achieved, when SSF was conducted at whole slurry solids loading of 15% with enzyme and yeast loading of 20 FPU/g glucan and 1.8 g/kg (wet), respectively, without nutrition supplementation and sterilization. It is believed that with mechanical mixing, enzyme loading can be reduced without reducing ethanol yield with extended fermentation duration.

Keywords: forest residue; pretreatment; liquefaction; enzymatic hydrolysis/saccharification; fermentation; high titer bioethanol; detoxification

---

1. Introduction

Fermentation of sugars from lignocelluloses has been proposed as a viable pathway for the production of renewable biofuels to supplement petro-fuels for sustainable economic development [1]. Feedstock is a major cost factor in producing cellulosic biofuels. Using low cost and underutilized feedstock such as harvest forest residues can be a winning proposition [2]. Forest residue can be sustainably produced in large quantities in various regions of the world [3,4], however, requires extensive pretreatment to remove its strong recalcitrance to enzymatic sugar production [5]. Severe pretreatments were often applied to remove this recalcitrance but resulted in the production of fermentation inhibitors [6,7]. Pretreatment optimization and strategies can substantially reduce inhibitor formation to facilitate high solids saccharification and fermentation without detoxification [8–12]. Proper management of inhibitor and sugar profiles in fermentation can improve ethanol productivity as xylose fermentation using *Saccharomyces cerevisiae* is especially sensitive to inhibitor profile [13,14]. Some studies suggest that initial sugar concentration can affect ethanol productivity [10,15]. Therefore, it is worthwhile to study the potential batch fermentation options to maximize ethanol production.

The objective of the present study is therefore to evaluate several batch fermentation approaches under various conditions for ethanol production from a softwood forest residue. The forest residue was pretreated by Sulfite Pretreatment to Overcome the Recalcitrance of Lignocelluloses (SPORL) [16]. SPORL was chosen for its robust performance in bioconversion of softwood biomass to sugars and biofuel [11,17]. Few processes have demonstrated good performance in effective removal of the recalcitrance of softwood biomass [5,18]. A SPORL pretreated Douglas-fir forest residue whole slurry from a previous study [10] at high SO_2 loading was used because of its low inhibitors, relatively short pretreatment residence time of 1 h, and potentially low metallurgy requirement due to the low pretreatment temperature of 140 °C. Separated enzymatic hydrolysis and fermentation (SHF), simultaneous enzymatic saccharification and fermentation (SSF), as well as quasi-simultaneous enzymatic saccharification and fermentation (Q-SSF) were evaluated. In Q-SSF, the pretreated whole slurry solids (WSS) was first liquefied through a pre-hydrolysis period at elevated temperature optimized for the cellulase used, i.e., 50 °C for the present study. Fermentation was then performed after cooling down the liquefied WSS to the desired temperature for the yeast, i.e., 38 °C in the present study. Ranges of liquefaction time, cellulase and yeast loadings were all evaluated in this study.

2. Materials and Methods

2.1. Materials

Douglas-fir forest residue was collected from a regeneration harvest in a primarily Douglas-fir stand in Lane County, OR, USA. The forest residue was ground and screened as described previously [9,19] to reduce bark and ash content and dead load in transportation. The accept forest residues were labelled as FS-10 and shipped to USDA Forest Products Laboratory, Madison, WI, after air drying to a moisture content of approximately 15%.

A commercial complex cellulase enzyme, Cellic® CTec3 (abbreviated CTec3), was complimentary provided by Novozymes North America (Franklinton, North Carolina, USA). The cellulase activity was 217 FPU/mL as calibrated by a literature method [20]. All the chemicals used were ACS reagent grade and purchased from Sigma-Aldrich (St. Louis, MO, USA).

Saccharomyces cerevisiae YRH400, an engineered fungal strain for xylose fermentation, was obtained from USDA Agriculture Research Service [21]. The strain was grown at 30 °C for 2 days on YPD agar plates containing 10 g/L yeast extract, 20 g/L peptone, 20 g/L glucose, and 20 g/L agar. A colony from the plate was transferred by loop to 50 mL liquid YPD medium in a 125 mL flask and cultured for overnight at 30 °C with agitation at 150 rpm on a shaking bed incubator (Thermo Fisher Scientific, model 4450, Waltham, MA, USA). The biomass concentration was monitored using optical density at 600 nm (OD_{600}) by a UV-vis spectrophotometer (Model 8453, Agilent Technologies, Palo Alto, CA, USA). The yeast seed at logarithmic phase with an average OD_{600} of approximately 18 was harvested and centrifuged at 5000 rpm, wet solid pellet was collected for fermentation.

2.2. Pretreatment

SPORL pretreatment of Douglas fir in a pilot scale wood pulping digester of 390-L was described previously [10]. The same pretreated FS-10 whole slurry sample labeled as C-t60 was used in this study. The C-t60 was produced by pretreating FS-10 at 140 °C with a liquor to wood ratio (L/W) of 4:1 (L/kg) and a targeted total SO_2 concentration of 80 g/L and a combined (with magnesium oxide) SO_2 loading of 11 g/L in the pretreatment liquor. These chemical loadings were chosen to simulate the chemistry in a commercial sulfite pulp mill in the US that recovers magnesium in operation. The pretreatment lasted 60 min after 32 min of ramping to 140 °C. Liquor circulation provided good mixing in reaction.

The collected solids and neutralized liquor were disk milled together. The measured whole slurries had a total solids content of 21.6%.

2.3. Enzymatic Hydrolysis

Enzymatic hydrolysis of the whole slurry solid (WSS) was carried out in 250 mL Erlenmeyer on a shaking incubator (Thermo Fisher Scientific, Model 4450, Waltham, MA, USA) to evaluate the effect of cellulase loading on WSS saccharification efficiency. Solid calcium carbonate was used to adjust the pH of the whole slurry of total solids of 15% (*w/w*) to approximately 5.5. Elevated pH of 5.5 was used to reduce nonproductive cellulase binding to lignin [22,23]. Hydrolysis was conducted at 50 °C and incubator shaking frequency of 200 rpm for 72 h. CTec3 loadings were 5, 10, 20 FPU/g glucan based on glucan in water insoluble solids (WIS). Hydrolysates were sampled periodically. At the end of enzymatic hydrolysis, solid residues were separated by centrifugation at 13,000× *g* for 5 min. Hydrolysates were analyzed for glucose. Duplicate hydrolysis runs were performed. Means and standard deviations were used as error bars in plots.

2.4. Enzymatic Saccharification and Fermentation

Enzymatic saccharification and fermentation experiments of the SPORL whole slurry C-t60 were also carried out in 125 mL Erlenmeyer flasks on the same shaker incubator described in the previous section. The pH of the slurry was again adjusted to 5.5 using solid calcium carbonate. The enzymatic hydrolysis was conducted at 15% total solids loading (*w/w*) and with CTec3 loading of 20 FPU/g of glucan. Hydrolysis was carried out at 50 °C and 200 rpm for 72 h, 8 h, and 0 h, respectively, before subsequent separate saccharification and fermentation (SHF), quasi-simultaneous saccharification and fermentation (Q-SSF), and true simultaneous saccharification and fermentation (SSF). No sterilization was applied prior to all these fermentation runs. The cultured yeast broth with OD_{600} = 18 was directly applied to inoculate the enzyme loaded pretreated whole slurry at volumetric loadings of 6, 8, and 10%, corresponding to wet yeast pellet (obtained by centrifuging at 5000 rpm for 20 min) of 1.3, 1.8, and 2.2 g/kg, respectively. The hydrolyzed or liquefied whole slurries were cooled down to 30 °C for SHF and 38 °C for Q-SSF and SSF.

Different enzyme loadings of 5, 10 and 20 FPU/g of glucan in WIS were used to evaluate the effect of enzyme dosage on ethanol production in SSF without pre-hydrolysis for liquefaction of solids.

Control fermentation runs were conducted in SSF mode at CTec3 loading of 20 FPU/g and yeast volumetric loading of 8% without sterilization but with the supplementation of nutrients: Yeast extract = 5 g/L, (NH4)₂SO₄ = 2 g/L, NaH₂PO₄ 5 = g/L. The effect of without nutrient supplementation and the effect of sterilization were also studied. The whole slurry C-t60 was autoclaved at 121 °C for 15 min prior to SSF fermentation for sterilization. The results were compared with the control run without sterilization.

All fermentation runs were carried out in duplicate. Samples were withdrawn at 4, 12, 24, 48, 72, 96, and 120 h, and centrifuged at 13000× *g* for 5 min. The supernatants were used for sugar and ethanol analyses. Furan concentrations were undetectable. Replicate analyses were conducted and the means and standard deviations of the duplicate fermentation runs were reported. Reported ethanol yield in percent theoretical was calculated based on the glucan, mannan, and xylan content in the untreated *Douglas-fir* residue FS-10 as expressed in the following equation:

$$y_{EtOH\ Theoretical}\ (\%) = \frac{C_{EtOH} \times V_{Broth}}{0.511[\frac{C_{glu}+C_{Man}}{0.9} + \frac{C_{Xyl}}{0.88}]} \times \frac{Y_{Pretrta\ Total\ Solids}}{m_{Pretreated\ Total\ Solids}} \times 100 \tag{1}$$

where C_{EtOH} *and* V_{Broth} are the terminal ethanol concentration (g/L) and the volume (L) of the fermentation broth; C_{Glu}, C_{Man}, C_{Xyl} are glucan, mannan, and xylan content (g/kg), respectively, in untreated woody biomass; $Y_{Pretreated\ Total\ Solids}$ is the yield of total solids (in oven dry weight) from pretreating one kg of woody biomass (g/kg); $m_{Pretrta\ Total\ Solids}$ is the total solids in oven dry weight of the sample used in the fermentation experiment (g).

2.5. Analytical Methods

Untreated and SPORL-pretreated FS-10 samples were first Wiley milled (Model No. 2; Arthur Thomas Co, Philadelphia, PA, USA) to 20 mesh and then hydrolyzed in two stages using sulfuric acid of 72% (*v/v*) at 30°C for 1 h and 3.6% (*v/v*) at 120°C for 1 h, respectively, as described previously [24]. Carbohydrates of the acid hydrolysates were analyzed by high performance anion exchange chromatography with pulsed amperometric detection (Dionex ICS-5000, Thermo Scientific, Sunnyvale, CA, USA). Klason lignin was quantified gravimetrically.

The enzymatic hydrolysates and fermentation broths were analyzed for monosaccharides, ethanol, furans, and organic acids as described previously [10] using two Dionex HPLC systems (Ultimate 3000, Thermo Scientific, Sunnyvale, CA, USA) equipped with a RI (RI-101) and a UV (VWD-3400RS) detector, respectively. For fast analysis, glucose in the enzymatic hydrolysates was measured using a commercial glucose analyzer (YSI 2700S, YSI Inc., Yellow Springs, OH, USA).

3. Results and Discussion

3.1. Compositional Analysis of SPORL Pretreated Douglas-Fir Residue

The chemical composition of the untreated FS-10 is listed in Table 1 along with component recoveries from the SPORL pretreated washed water insoluble solids (WIS) and sugar and degradation products concentrations in the pretreatment spent liquor. Total of 186.88 kg spent liquor were collected which represent a loss of water of 13.12 kg (200 kg of total water was used including moisture in the pretreatment with liquor to wood ratio of 4:1 L/kg) due to hot blow of the pretreated materials in the reactor at the end of pretreatment.

Table 1. Chemical composition of untreated and SPORL-pretreated unwashed wet solids and freely drainable spent liquor and total component recovery from the whole slurry solids.

	Untreated FS-10	Pretreated FS-10 Wet Solids	Freely Drainable Liquor [1]	Total Recovery (%)
Wet weight (kg)	66.41	126.50	94.35	88.1
Oven dry solids (kg)	50	43.97	7.25	101.0 [2]
Klason lignin (kg)	14.65	8.67	5.27	95.2
Arabinan (kg)	0.51	0.32		
Galatan (kg)	1.00	1.13		
Glucan (kg)	20.49	17.27	0.63 (6.72)	87.4
Mannan (kg)	4.84	3.57	1.67 (17.66)	108.3
Xylan (kg)	2.85	1.76	0.78 (8.29)	89.1

[1] numbers in parenthesis are component concentration in g/L; [2] calculated including the amount of 0.607 kg of Mg(OH)$_2$ applied.

3.2. Evaluation of Pretreatment Effectiveness and Effect of Enzyme Loading on Saccharification

Enzymatic hydrolysis of whole slurry solids (WSS) was conducted to evaluate the effectiveness of SPORL pretreatment for saccharification. Time-dependent substrate enzymatic digestibility (SED) shown in Figure 1 was defined as the percentage of glucan in WSS enzymatically hydrolyzed to glucose. Glucose release with 20 FPU/g glucan CTec3 loading increased sharply with time and reached over 90% within 24h with a glucose titer of 65 g/L and then increased slowly and reached 93% in 72 h with a titer of 67.7 g/L. The glucose release at the end of 72 h hydrolysis were 76.7 and 59.6% with a titer of 55.4 and 43.1 g/L at 48 h at 10 and 5 FPU/g glucan loadings, respectively. These results suggested SPORL pretreatment was effective for maximal saccharification of *Douglas-fir* residue FS-10, in agreement with a previous study that evaluate the water insoluble solids (WIS) of the same pretreated FS-10 [10]. Even with a 4-fold reduction in cellulase loading from 20 to 5 FPU/g glucan, the enzymatic saccharification efficiency remained at approximately 60%. Most of the fermentation experiments, nevertheless, were carried out using cellulase loading of 20 FPU/g glucan to facilitate liquefaction because mixing was done by shaking on a bed without mechanical mixing.

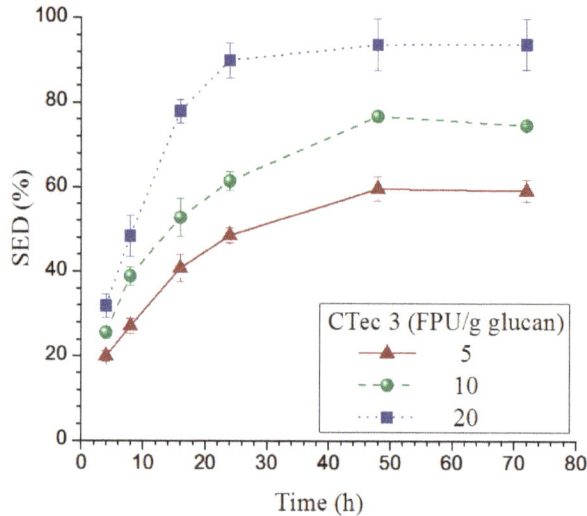

Figure 1. Time-dependent substrate enzymatic digestibility of SPORL-pretreated FS-10 WSS at 15% total solids and three CTec3 loadings.

3.3. Comprison of Three Different Fermentation Options on Ethanol Production

Ethanol yield in a traditional SHF was low, probably due to end-product inhibition of the hydrolysis. Glucose inhibition becomes important when glucose concentration reaches over 50 g/L [25]. The concept of performing SSF was proposed in the 1970s [26] to eliminate end product inhibition. Furthermore, there are several additional potential advantages for using SSF, the combination of hydrolysis and fermentation decreases the number of vessels needed and, therefore investment costs. The decrease in capital investment has been estimated to be greater than 20%. This is quite important since the capital costs can be expected to be comparable to the raw material costs in ethanol production from lignocelluloses [27]. In Q-SSF, lignocellulosic slurry was pre-hydrolyzed using enzymes to liquefy solids to produce a certain amount of monomeric and oligomeric sugars prior to fermentation using microorganisms. This approach avoids the disadvantage of low temperature hydrolysis required for yeast survival during liquefaction. The elevated liquefaction temperature, not only improved enzymatic activity, but also reduced the viscosity of the hydrolysate, which facilitates mixing for saccharification and fermentation.

Three fermentation strategies of SHF, Q-SSF and SSF of C-t60 WSS were carried at 15% (*w*/*w*) total solids loading with the supplementation of nutrient. The whole slurry was pre-hydrolyzed for 72 h, 8 h, and 0 h at 50 °C and 200 rpm on a shaking bed, which resulted in glucose concentrations of 67.8, 32.1 and 10 g/L, respectively. The yeast loading was 8% *v*/*v*. The results showed both Q-SSF and SSF produced a higher terminal ethanol concentration than SHF (Figure 2a). An obvious lagging phase in ethanol production can be clearly observed in SHF in the first 12 h, probably due to the high glucose concentration, which resulted in a terminal ethanol concentration of 38.6 g/L at 96 h. Compared with SHF, no visible lagging phase was observed in both Q-SFF and SFF, and ethanol production started immediately after inoculation though slower than SHF between 12 and 24 h, perhaps due to lower glucose concentration. In the first 36 h, Q-SSF had higher ethanol productivity than SSF perhaps due to the availability of more glucose and the reduced viscosity in the Q-SSF slurry due to pre-hydrolysis and liquefaction. However, Q-SSF and SSF reached the same terminal ethanol concentration of 42.3 g/L at 72 h. Glucose concentration continued to increase for the SSF and Q-SSF runs even at the end of 96 h, suggesting continued saccharification and better managing yeast activity or adding more yeast at this stage could further increase ethanol yield. Continued saccharification at the end of fermentation was also observed in a previous study [11].

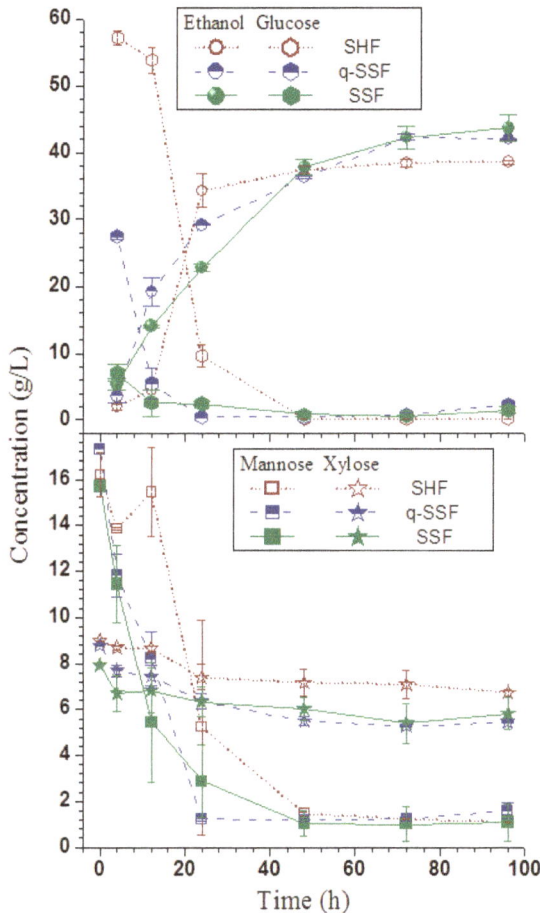

Figure 2. Comparisons of time-dependent fermentation results of SPORL pretreated *Douglas-fir* residue FS-10 among different fermentation options at CTec 3 of 20 FPU/g glucan and 8% (*v/v*) yeast loading. (**a**) Ethanol and glucose concentrations; (**b**) Mannose and xylose concentrations.

Time-dependent hemicellulosic sugar consumptions are shown in Figure 2b. Mannose was consumed rapidly by YRH400 in the first 48 h, especially in the SSF and Q-SSF runs. This is probably because glucose concentration was lower in these two fermentations (Figure 2a) than that in the SHF run, which resulted in better utilization of mannose and even xylose to some extent. After 48 h, glucose concentrations approaches zero and hemicellulosic sugar consumptions were pretty much completed in all three runs (Figure 2a,b). It appears that the yeast was starved and under stress, as discussed above, and incapable of consuming hemicellulosic sugars at very low concentrations. The final utilization of mannose in three fermentation runs, however, were very high, over 90%. The final xylose consumption was 38% and 32% in q-SFF and SSF, respectively, higher than the 25% in SHF.

3.4. Comprison of Different Enzyme Loadings in SSF Fermentation

The results of enzymatic hydrolysis of WSS at different enzyme loadings indicated insufficient cellulose saccharification at 5 and 10 FPU/g glucan at 15% loading of WSS (Figure 1). However, enzyme is a very important cost factor in bioethanol production [27]. SSF of WSS was carried out at enzyme loadings of 5, 10 and 20 FPU/g glucan to evaluate the effect of enzyme loading on ethanol production. The results shown in Figure 3a clearly indicated a strong positive correlation between enzyme loading and terminal ethanol yield. Ethanol concentration from 20 FPU/g glucan was much higher than 5 and 10 FPU/g glucan. It is interesting to notice that glucose concentration was highest at the highest enzyme loading of 20 FPU/g glucan during 4 to 24 h, followed by loading of 10 FPU/g glucan, and lowest at 5 FPU/g glucan. This perhaps suggests rapid cellulose saccharification at high cellulase loadings and initial (0–12 h) fermentation may be slower (Figure 2a). Glucose concentration was increased after 96 h at enzyme loadings of 10 FPU/g glucan (Figure 3a) similar to that observed in Figure 2a at 20 FPU/g glucan and in a previous study [11]. This observation indicates that better managing yeast activity can improve ethanol yield. Furthermore, ethanol concentration was not plateaued at the two low CTec 3 loadings of 5 and 10 FPU/g glucan (Figure 3a), suggesting extending fermentation time can improve ethanol yield even at low cellulase loadings. One can balance capital and enzyme costs by using low cellulase loadings with extended fermentation time.

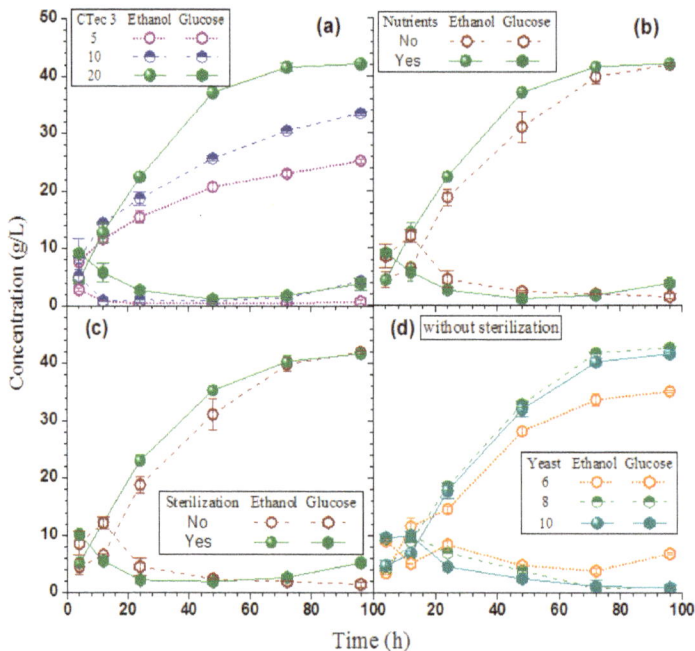

Figure 3. Effects of (**a**) cellulase loading; (**b**) nutrient supplementation; (**c**) sterilization, and (**d**) yeast loadings on time-dependent fermentation results of SPORL pretreated Douglas fir residue FS-10 WSS.

3.5. Effect of Nutrient Supplementation on the Ethanol Production

Fermentation runs using SSF were carried out with and without the supplementation of nutrients at CTec3 loading of 20 FPU/g glucan without sterilization. Ethanol production with nutrients supplementation was higher than that without before 96 h (Figure 3b), probably because the yeast needed more time to acclimatize in the media without nutrients. Terminal ethanol production at 96 h, however, was the same (43 g/L) with and without nutrients.

3.6. Effect of Sterilization

In general, the SPORL pretreated whole slurry C-t60 was quite sterile. However the pretreated FS-10 sample has been stored in a cold room for quite a few months. It is worth studying whether or not sterilization help ethanol productivity. The SSF results clearly showed a lagging phase within the first 12 h for the fermentation run without sterilization (Figure 3c). Glucose consumption was much faster when the sample was autoclaved at 121 °C for 15 min. This perhaps due to the fact that growth of *Saccharomyces cerevisiae* YRH400 was suppressed by other microorganisms that existed in the sample. After 48 h, these microorganisms may not be able to survive due to elevated ethanol concentrations. The inoculated strain YRH400 became the dominant strain. As a result, terminal ethanol concentration at 96 h was not affected due to the lack of sterilization.

3.7. Effect of Yeast Loading on the Ethanol Production

In industrial practice, enzyme and yeast cell concentrations should be appropriately balanced to minimize the costs for yeast and enzyme production. A higher yeast concentration in the SSF can result in a lower overall ethanol yield when the cost for yeast production is taken into account [28]. However, lowering the yeast concentration may low the volumetric productivity, and lead to a stuck fermentation. Three different yeast volumetric loadings of 6%, 8%, and 10% were inoculated in SSF with (not shown) and without (Figure 3d) sterilization. With sterilization, glucose consumption and the ethanol production were slightly slower at a lower yeast loading of 6 v/v % in the first 24 h, the ethanol production, however, reached the same concentration of approximately 43 g/L at the two higher yeast loadings after 96 h fermentation. These results are in agreement with a study from the literature using SO_2 steam explosion pretreated spruce [28]. When sterilization was not applied, however, glucose consumption and ethanol production were obviously slower at the lowest volumetric yeast loading of 6% than the two higher loadings of 8% and 10%. The final ethanol concentration at 96 h was 35 g/L at the lower yeast loading of 6%, substantially lower than the approximately 42 g/L at the two higher yeast loadings. This suggests sterilization can suppress other microorganism growth, which allows low yeast loading in ethanol fermentation.

4. Conclusions

The whole slurry solids (WSS) obtained from SPORL-pretreated *Douglas-fir* was evaluated for the production of bioethanol using *Saccharomyces cerevisiae* YRH400. The production of ethanol was investigated in a batch fermentation using modes, i.e., separate hydrolysis and fermentation (SHF), Quasi-simultaneous saccharification and fermentation (Q-SSF), and simultaneous saccharification and fermentation (SSF). Under SSF, the effects of different enzyme and yeast loadings, nutrient supplementation, and sterilization of the substrate on ethanol production were investigated. A final ethanol titer of 43.2 g/L was achieved when SSF was conducted at whole slurry solids loading of 15%, enzyme and yeast loading of 20 FPU/g glucan and 1.8 g/kg yeast loading (wet cell), respectively, with sterilization but without nutrition supplementation.

Acknowledgments: The authors would like to acknowledge the financial support of the Agriculture and Food Research Initiative (AFRI) Competitive grant (No. 2011-68005-30416), USDA National Institute of Food and Agriculture (NIFA) through the Northwest Advanced Renewables Alliance (NARA). The authors also appreciate Novozymes North America for providing the CTec3 enzymes; Fred Matt of USFS-FPL for conducting the chemical composition analysis of forest residue samples; Bruce Dien and Ron Hector of USDA-ARS for providing the

YRH-400 strain. Rolland Gleisner (USDA Forest Products Lab (FPL)) and William Gilles (Formerly with FPL) conducted pilot scale SPORL pretreatment. Funding was also received from The Chinese Scholarship Council for the visiting appointment of Mingyan Yang at the USDA Forest Products Lab.

Author Contributions: Mingyan Yang and J.Y. Zhu conceived and designed the experiments; Mingyan Yang conducted the experiments; Hairui Ji conducted analytical work and calculations. Mingyan Yang and J.Y. Zhu analyzed the results and wrote the paper.

Conflicts of Interest: J.Y. Zhu is a co-inventor of the SPORL technology. The founding sponsors had no role in the design of the study; in the collection, analyses, or interpretation of data; in the writing of the manuscript, and in the decision to publish the results.

References

1. U.S. DOE. Breaking the Biological Barriers to Cellulosic Ethanol: A Joint Research Agenda. *A Research Road Map Resulting from the Biomass to Biofuel Workshop Sponsored by The Department of Energy*; U.S. Department of Energy: Rockville, MD, USA, 2005.

2. National Research Council. *Renewable Fuel Standard: Potential Economic and Environmental Effects of US Biofuel Policy*; The Natinal Academies Press: Washington, WA, USA, 2011.

3. Gan, J.; Smith, C.T. Availability of logging residues and potential for electricity production and carbon displacement in the USA. *Biomass Bioenerg.* **2006**, *30*, 1011–1020. [CrossRef]

4. Perlack, R.D.; Stokes, B.J.; DOE. *2011. U.S. Billion-Ton Update: Biomass Supply for a Bioenergy and Bioproducts Industry*; Oakridge National Laboratory: Oak Ridge, TN, USA, 2011.

5. Zhu, J.Y.; Pan, X.J. Woody Biomass Pretreatment for Cellulosic Ethanol Production: Technology and Energy Consumption Evaluation. *Bioresour. Technol.* **2010**, *101*, 4992–5002. [CrossRef] [PubMed]

6. Larsson, S.; Palmqvist, E.; Hahn-Hagerdal, B.; Tengborg, C.; Stenberg, K.; Zacchi, G.; Nilvebrant, N.-O. The generation of fermentation inhibitors during dilute acid hydrolysis of softwood. *Enzyme Microb. Technol.* **1999**, *24*, 151–159. [CrossRef]

7. Zhou, H.; Leu, S.-Y.; Wu, X.; Zhu, J.Y.; Gleisner, R.; Yang, D.; Qiu, X.; Horn, E. Comparisons of high titer ethanol production and lignosulfonate properties by SPORL pretreatment of lodgepole pine at two temperatures. *RSC Adv.* **2014**, *4*, 27033–27038. [CrossRef]

8. Zhang, C.; Houtman, C.J.; Zhu, J.Y. Using low temperature to balance enzymatic saccharification and furan formation in SPORL pretreatment of Douglas-fir. *Process Biochem.* **2014**, *49*, 466–473. [CrossRef]

9. Cheng, J.; Leu, S.-Y.; Zhu, J.Y.; Gleisner, R. High titer and yield ethanol production from undetoxified whole slurry of Douglas-fir forest residue using pH-profiling in SPORL. *Biotechnol. Biofuels* **2015**, *8*, 22. [CrossRef] [PubMed]

10. Gu, F.; Gilles, W.; Gleisner, R.; Zhu, J.Y. Fermentative high titer ethanol production from a Douglas-fir forest residue without detoxification using SPORL: High SO$_2$ loading at a low temperature. *Ind. Biotechnol.* **2016**, *12*, 168–175. [CrossRef]

11. Zhu, J.Y.; Chandra, M.S.; Gu, F.; Gleisner, R.; Reiner, R.; Sessions, J.; Marrs, G.; Gao, J.; Anderson, D. Using sulfite chemistry for robust bioconversion of Douglas-fir forest residue to bioethanol at high titer and lignosulfonate: A pilot-scale evaluation. *Bioresour. Technol.* **2015**, *179*, 390–397. [CrossRef] [PubMed]

12. Zhou, H.; Zhu, J.Y.; Gleisner, R.; Qiu, X.; Horn, E.; Negron, J. Pilot-scale demonstration of SPORL for bioconversion of lodgepole pine to bio-ethanol and lignosulfonate. *Holzforschung* **2016**, *70*, 21–30.

13. Zhou, H.; Lan, T.; Dien, B.S.; Hector, R.E.; Zhu, J.Y. Comparisons of Five *Saccharomyces cerevisiae* strains for Ethanol Production from SPORL Pretreated Lodgepole Pine. *Biotechnol. Prog.* **2014**, *30*, 1076–1083. [CrossRef] [PubMed]

14. Almeida, J.R.M.; Runquist, D.; Sànchez Nogué, V.; Lidén, G.; Gorwa-Grauslund, M.F. Stress-related challenges in pentose fermentation to ethanol by the yeast Saccharomyces cerevisiae. *Biotechnol. J.* **2011**, *6*, 286–299. [CrossRef] [PubMed]

15. Hoyer, K.; Galbe, M.; Zacchi, G. The effect of prehydrolysis and improved mixing on high-solids batch simultaneous saccharification and fermentation of spruce to ethanol. *Process Biochem.* **2013**, *48*, 289–293. [CrossRef]

16. Zhu, J.Y.; Pan, X.J.; Wang, G.S.; Gleisner, R. Sulfite pretreatment (SPORL) for robust enzymatic saccharification of spruce and red pine. *Bioresour. Technol.* **2009**, *100*, 2411–2418. [CrossRef] [PubMed]

17. Leu, S.-Y.; Gleisner, R.; Zhu, J.Y.; Sessions, J.; Marrs, G. Robust Enzymatic Saccharification of a Douglas-fir Forest Harvest Residue by SPORL. *Biomass Bioenerg.* **2013**, *59*, 393–401. [CrossRef]

18. Yamamoto, M.; Niskanen, T.; Iakovlev, M.; Ojamo, H.; van Heiningen, A. The effect of bark on sulfur dioxide-ethanol-water fractionation and enzymatic hydrolysis of forest biomass. *Bioresour. Technol.* **2014**, *167*, 390–397. [CrossRef] [PubMed]

19. Zhang, C.; Zhu, J.Y.; Gleisner, R.; Sessions, J. Fractionation of Forest Residues of Douglas-fir for Fermentable Sugar Production by SPORL Pretreatment. *Bioenerg. Res.* **2012**, *5*, 978–988. [CrossRef]

20. Wood, T.M.; Bhat, M. Methods for Measuring Cellulase Activities. In *Methods in Enzymology*; Colowick, S.P., Kaplan, N.O., Eds.; Academic Press, Inc.: New York, NY, USA, 1988; pp. 87–112.

21. Hector, R.E.; Dien, B.S.; Cotta, M.A.; Qureshi, N. Engineering industrial Saccharomyces cerevisiae strains for xylose fermentation and comparison for switchgrass conversion. *J. Ind. Microbio. Biotechnol.* **2011**, *38*, 1193–1202. [CrossRef] [PubMed]

22. Lan, T.Q.; Lou, H.; Zhu, J.Y. Enzymatic saccharification of lignocelluloses should be conducted at elevated pH 5.2–6.2. *Bioenerg. Res.* **2013**, *6*, 476–485. [CrossRef]

23. Lou, H.; Zhu, J.Y.; Lan, T.Q.; Lai, H.; Qiu, X. pH-induced lignin surface modification to reduce nonspecific cellulase binding and enhance enzymatic saccharification of lignocelluloses. *ChemSusChem* **2013**, *6*, 919–927. [CrossRef] [PubMed]

24. Luo, X.; Gleisner, R.; Tian, S.; Negron, J.; Horn, E.; Pan, X.J.; Zhu, J.Y. Evaluation of mountain beetle infested lodgepole pine for cellulosic ethanol production by SPORL pretreatment. *Ind. Eng. Chem. Res.* **2010**, *49*, 8258–8266. [CrossRef]

25. Holtzapple, M.; Cognata, M.; Shu, Y.; Hendrickson, C. Inhibition of *Trichoderma reesei* cellulase by sugars and solvents. *Biotechnol. Bioeng.* **1990**, *36*, 275–287. [CrossRef] [PubMed]

26. Gauss, W.F.; Suzuki, S.; Takagi, M. Manufacture of Alcohol from Cellulosic Materials Using Plural Ferments. U.S. Patent 3990944 A, 9 November 1976.

27. Olofsson, K.; Bertilsson, M.; Lidén, G. A short review on SSF—An interesting process option for ethanol production from lignocellulosic feedstocks. *Biotechnol. Biofuels* **2008**, *1*, 7. [CrossRef] [PubMed]

28. Sassner, P.; Galbe, M.; Zacchi, G. Bioethanol production based on simultaneous saccharification and fermentation of steam-pretreated Salix at high dry-matter content. *Enzyme Microb. Technol.* **2006**, *39*, 756–762. [CrossRef]

fermentation

MDPI

Review
Yeast Nanobiotechnology

Ronnie Willaert [1,*], Sandor Kasas [2], Bart Devreese [3] and Giovanni Dietler [2]

[1] IJRG VUB-EPFL, NanoBiotechnology & NanoMedicine (NANO), Alliance Research Group VUB-UGent NanoMicrobiology (NAMI), Vrije Universiteit Brussel, Brussels 1050, Belgium
[2] IJRG VUB-EPFL, NanoBiotechnology & NanoMedicine (NANO), Laboratory of the Physics of Living Matter, Ecole Polytechnique de Lausanne, Lausanne 1015, Switzerland; Sandor.Kasas@epfl.ch (S.K.); Giovanni.Dietler@epfl.ch (G.D.)
[3] IJRG VUB-EPFL, NanoBiotechnology & NanoMedicine (NANO), Alliance Research Group VUB-UGent NanoMicrobiology (NAMI), Laboratory for Protein Biochemistry and Biomolecular Engineering (L-ProBE), Ghent University, Ghent 9000, Belgium; Bart.Devreese@UGent.be
* Correspondence: Ronnie.Willaert@vub.ac.be; Tel.: +32-2-629-1846

Academic Editor: Badal C. Saha
Received: 6 August 2016; Accepted: 13 October 2016; Published: 21 October 2016

Abstract: Yeast nanobiotechnology is a recent field where nanotechniques are used to manipulate and analyse yeast cells and cell constituents at the nanoscale. The aim of this review is to give an overview and discuss nanobiotechnological analysis and manipulation techniques that have been particularly applied to yeast cells. These techniques have mostly been applied to the model yeasts *Saccharomyces cerevisiae* and *Schizosaccaromyces pombe*, and the pathogenic model yeast *Candida albicans*. Nanoscale imaging techniques, such as Atomic Force Microscopy (AFM), super-resolution fluorescence microscopy, and electron microscopy (scanning electron microscopy (SEM), transmission electron microscopy (TEM), including electron tomography) are reviewed and discussed. Other nano-analysis methods include single-molecule and single-cell force spectroscopy and the AFM-cantilever-based nanomotion analysis of living cells. Next, an overview is given on nano/microtechniques to pattern and manipulate yeast cells. Finally, direct contact cell manipulation methods, such as AFM-based single cell manipulation and micropipette manipulation of yeast cells, as well as non-contact cell manipulation techniques, such as optical, electrical, and magnetic cells manipulation methods are reviewed.

Keywords: yeasts; Atomic Force Microscopy (AFM); super-resolution fluorescence microscopy; electron microscopy; force spectroscopy; nanomotion analysis; yeast cell patterning; non- and direct-contact cell manipulation; optical/magnetic tweezer; nanoscale imaging

1. Introduction

Nanotechnology is the ability to work at the atomic, molecular, and supramolecular levels (on the scale of ~1–100 nm) to understand, create, and use material structures, devices, and systems with fundamentally new properties and functions resulting from their small structure [1]. Nanobiotechnology is defined as a field that applies nanoscale principles and techniques to understand and transform biosystems (living or nonliving) and that uses biological principles and materials to create new devices and systems integrated from the nanoscale [2]. The biological and physical sciences share a common interest in small structures (the definition of "small" depends on the application, but can range from 1 nm to 1 mm) [3]. A bacterial cell is approximately 1 μm, a yeast cell 5 μm, and a mammalian cell is 10 μm when rounded and 50 μm when fully spread in attached culture. A vigorous trade across the borders of these areas of science is developing around new materials and tools (largely from the physical sciences) and new phenomena (largely from the biological sciences). The physical sciences offer tools for the synthesis and fabrication of devices for measuring the characteristics of cells and sub-cellular components and of materials useful in cell and molecular biology. Biology offers a window into the most sophisticated collection of functional nanostructures that exist. Nanobiotechnology offers new solutions for the transformation of biosystems, and provides a broad technological platform for applications in several areas—including bioprocessing in industry, molecular medicine, investigating the health effects of nanostructures in the environment, improving food products (food conservation), and improving human performance [2].

This review discusses nanobiotechnological analysis and manipulation techniques that have been especially applied to yeast cells. Nanoscale imaging methods that allow imaging at nanometer resolution are reviewed: atomic force microscopy (AFM), super-resolution fluorescence microscopy, and electron microscopy (including scanning electron microscopy (SEM), transmission electron microscopy (TEM), and electron tomography). Force spectroscopy for the analysis of single biomolecule interactions or unfolding on live yeast cells, as well as single-cell force spectroscopy, and the recently developed AFM-based nanomotion analysis of cells is reviewed and discussed. Since single-cell analysis has increasingly been recognised as the key technology for the elucidation of cellular functions which are not accessible from bulk measurements of the population level, nano/micro single-yeast cell manipulation techniques are reviewed. Yeast cell patterning techniques, such as microcontact printing, mechanical cell patterning, and the use of robotic cell printing are discussed. Finally, direct-contact (such as AFM-based and micropipette-based) and non-contact (such as optical, electrical, and magnetic cell) yeast cell manipulation techniques are reviewed.

2. Yeast Nanobiotechnological Analyses

2.1. Nanoscale Imaging

We use microscopy in order to see objects in more detail. The best distance that one can resolve with optical instruments (disregarding all aberrations) is about 0.5 times the wavelength of light, or the order of 250 nm with visible radiation. High-resolution microscopy techniques that are used for nanoimaging and nanoscale characterisation have been developed in the last 20 years. They can be divided into three categories: optical microscopes, scanning probe microscopes (SPMs), and electron microscopes. Recently-developed microscopy-based technologies can also be used to control and manipulate objects at the nanoscale—i.e., single-cell as well as single-molecule manipulation and analysis.

2.1.1. Atomic Force Microscopy

Scanning probe microscopes (SPMs) are a family of instruments that are used to measure surface properties, and include atomic force microscopes (AFMs) and scanning tunneling microscopes (STMs). The main feature that all SPMs have in common is that the measurements are performed with a sharp probe operating in the near field; that is, scanning over the surface while maintaining a very close spacing to the surface. The STM—invented in the early nineteen-eighties by Binnig and Rohrer [4]—was the first to produce real-space images of atomic arrangements on flat surfaces. The development of the STM arose from an interest in the study of the electrical properties of thin insulating layers. This led to an apparatus in which the probe–surface separation was monitored by measuring electron tunneling between a conducting surface and a conducting probe. A few years later, Binnig and colleagues [5] announced the birth of the second member of the SPM family, the atomic force microscope (also known as the scanning force microscope, SFM). Numerous variations of these techniques have been developed since.

AFM is extensively used for imaging surfaces ranging from micro- to nanometer scales, with the objective of visualising and characterising surface textures and shapes [6]. It has evolved into an imaging method that yields structural details of biological samples, such as proteins, nucleic acids, membranes, and cells in their native environment. AFM is a unique technique for providing sub-nanometer resolution at a reasonable signal-to-noise ratio under physiological conditions.

It complements electron microscopy (EM) by allowing the visualisation of biological samples in buffers that preserve their native structure over extended time periods. Unlike EM, AFM yields 3D maps with an exceptionally good vertical resolution (less than a nanometer). Additionally, the measurement of mechanical forces at the molecular level provides detailed insights into the function and structure of biomolecular systems. Inter- and intramolecular interactions can be studied directly at the molecular level. Recently, improvements in the temporal resolution were made by the development of high-speed AFM [7]. This technique is capable of observing structure dynamics and dynamic processes at the sub-second to sub-100 ms temporal resolution and 2 nm lateral and 0.1 nm vertical resolution.

Since AFM imaging can be performed in physiological conditions, high resolution imaging of the yeast cell surface can be performed on living cells. Therefore, an appropriate cell immobilisation method has to be used that avoids cell detachment by the scanning probe. Several methods have been developed and used to perform high-resolution live yeast cell imaging and analysis (i.e., force spectroscopy). Yeast cells can be trapped in the pores of a filter membrane [8] (Figure 1A), in a hydrogel [9,10], or in microfabricated microwells [11,12]. Especially the model yeast *Saccharomyces cerevisiae* has been imaged at the nanoscale (Table 1). The cell walls of other yeasts have also been visualised: the model pathogenic yeast *Candida albicans*, and the other model yeast *Schizosaccharomyces pombe* (Table 1). Imaging can be easily combined with nanoindentation experiments to map the elasticity of the cell surface [13].

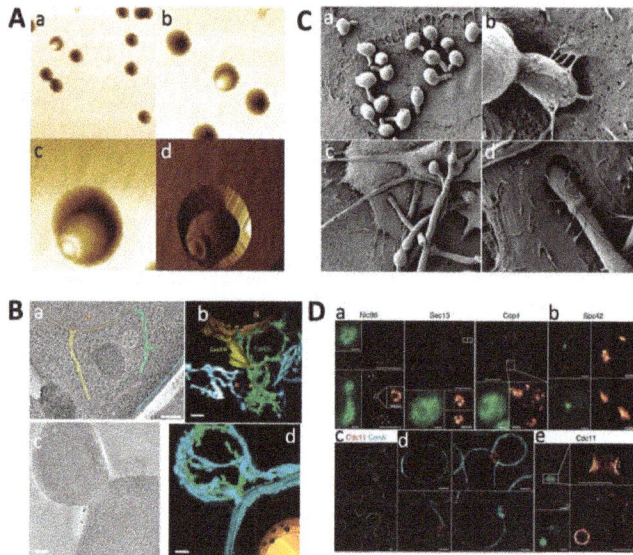

Figure 1. (**A**) Entrapment of a single *Saccharomyces cerevisiae* cell in the pore of a filter membrane. Atomic force microscopy (AFM) height (**a–c**) and amplitude (**d**) images. Courtesy of Dr. Ronnie Willaert, Vrije Universiteit Brussel, Belgium. (**B**) Electron tomography imaging of *S. cerevisiae*: 3D structural analysis of endoplasmic reticulum (ER) morphology. (**a,b**) 2D tomograph derived from a 200-nm-thick section shows the nuclear envelope (NE) (orange), plasma membrane ER (pmaER), central cisternal ER, tubular ER, and Golgi (pink; a) and the corresponding 3D model of (a) shows all ER domains in a WT yeast cell. The blue shade is the plasma membrane (PM); N is the nucleus; black holes on the NE are nuclear pores; (**c**) 2D tomograph of a mutant cell with a bud; (**d**) 3D model of ER domain organisation (the cytoplasmic face (cyto) of pmaER in blue and PM face of pmaER in red). Reprinted from ref. [14]. (**C**) Scanning electron microscopy (SEM) images of *Candida albicans* cells interacting with pharyngeal FaDu cells. (**a,b**) After 30 min of contact, the formation of germ tubes is visible. The cells attach to the FaDu cells through microvilli structures; (**c,d**) After 3 h of contact, the *C. albicans* cells produce long filamentous cells (hyphae and pseudohyphae), which penetrate FaDu cells. Magnification (a) ×4000,

(b) ×25,000, (c) ×4000, (d) ×16,000. Courtesy of Dr. Ronnie Willaert, Vrije Universiteit Brussel, Belgium. (**D**) Super-resolution photoactivatable localization microscopy (PALM) imaging of proteins in budding yeast green fluorescent protein (GFP)-fusion construct library. (**a**) Reconstructed super-resolution images of Nic96-, Sec13-, and Cop1-GFP; (**b**) Images of the spindle pole body protein Spc42-GFP; (**c**) Dual-colour reconstructed images of yeast cells expressing Cdc11-GFP (red) and the cell wall (cyan); (**d**) Reconstructed dual-colour images visualising the different organisational stages of Cdc11-GFP structures during the cell cycle; (**e**) Reconstructed images of septin Cdc11-GFP, which localises to a characteristic hourglass-shaped and later ring-like structure around the mother–bud neck. Reprinted with permission from ref. [15].

Table 1. Examples of AFM imaging of yeast cell surfaces.

Yeast Type	AFM Analysis	Objective	Refs
C. albicans	Imaging, cell surface elasticity	Effect of antifungal caspofungin	[16]
	Imaging, cell elasticity	Imaging mode evaluation	[17]
	Imaging, force spectroscopy using concanavalin A-functionalised tips	Mapping of adhesive properties	[18]
Candida parapsilosis	Imaging, adhesion force	Surface morphological characterisation	[19]
S. cerevisiae	Imaging	Immobilisation method	[8]
	Imaging	Immobilisation method	[9]
	Imaging, force spectroscopy using concanavalin A-functionalised tips	Mapping cell wall polysaccharides	[20]
	Imaging, cell elasticity	Mapping of cell elasticity	[21]
	Imaging	Cell surface change on thermal and osmotic stress	[22]
	Imaging, motion analysis	Nanomechanical motion analysis	[23]
	Imaging	Effect of electromagnetic field and antifungal nystatin on the cell wall	[24]
	Imaging, cell elasticity	Immobilisation method	[10]
	Imaging	Immobilisation method	[11]
	Imaging, cell surface elasticity	Effect of antifungal caspofungin	[16]
Sc. pombe	Imaging	Cell surface change on thermal and osmotic stress	[22]

2.1.2. Light Microscopy

Since the earliest examination of cellular structures, observing cells using a light microscope has fascinated biologists. Being able to observe processes as they happen with the use of light microscopy adds a vital extra dimension to our understanding of cell behaviour and function [25]. Microscopy has evolved to provide not only quantitative images but also a significant capability to perturb structure–function relationships in cells. These advances have been especially useful in the study of a wide range of biological processes, including cell adhesion and migration [26].

Recent advances in fluorescence microscopy have allowed the imaging of structures at extremely high resolutions [27]. The past decade witnessed an explosion of fluorescence microscopy-based approaches to image protein dynamics and interactions [28]. For example, fluorescence recovery after photobleaching (FRAP) or photo-activation using photo-convertible fluorescent proteins to assay protein mobility and maturation in cells [29]; and Förster resonance energy transfer (FRET) to monitor physical intra- or intermolecular associations in space and time [30,31].

Despite the advantages of standard fluorescence microscopy, ultra-structural imaging is not possible, owing to a resolution limit set by the diffraction of light (Rayleigh criterion) [32]. Therefore, the maximal spatial resolution of standard optical microscopy is around 200 nm. This limit is one to two orders of magnitude above the typical molecular length scales in cells. Several approaches have been used to break this diffraction limit (Table 2). The diffraction limit can be overcome by exploiting the distribution of fluorescence intensity from a single molecule. When imaged, a fluorophore behaves as a point source with an Airy disc point spread function. The center of mass of the function (and therefore the position of the molecule) can be obtained by performing a least-squares fit of an appropriate function (such as a Gaussian distribution) to the measured fluorescence intensity profile of the spot [33]. With a sufficient number of photons, these methods can provide a localisation of 1–2 nm (15 to 70 nm on intact cells), allowing the measurement of distances on the scale of individual proteins. Single-molecule detection offers new possibilities for obtaining sub-diffraction-limit spatial resolution [34–36].

Table 2. Super-resolution optical microscopy techniques (adapted from [37]).

Technique	Description	Spatial Resolution	Timescale
Fluorescence imaging with one-nanometer accuracy (FIONA)	Localises and tracks single-molecule emitters by finding the centre of their diffraction-limited point-spread function (PSF).	~1.5 nm	~0.3 ms
Single-molecule high-resolution colocalisation (SHREC)	Two-colour version of FIONA. Two fluorescent probes with different spectra are imaged separately and then localised and mapped onto the plane of the microscope.	<10 nm	~1 s per frame
Single-molecule high-resolution imaging with photobleaching (SHRImP)	Uses the strategy wherein, upon photobleaching of two or more closely-spaced identical fluorophores, their position is sequentially determined by FIONA, starting from the last bleached fluorophore.	~5 nm	~0.5 s per frame
Nanometer-localised multiple single-molecules (NALMS)	Uses a similar principle to single-molecule high-resolution imaging with photobleaching to measure distances between identical fluorescent probes that overlap within a diffraction-limited spot.	~8 nm	~1 s per frame
Photoactivatable localization microscopy (PALM)	Serially photoactivates and photodeactivates many sparse subsets of photoactivatable fluorophores to produce a sequence of images that are combined into a super-resolution composite.	~2 nm	~1 min
PALM with independently running acquisition (PALMIRA)	Records non-triggered spontaneous off–on–off cycles of photoswitchable fluorophores without synchronising the detector to reach faster acquisition.	~50 nm	~2.5 min
Single particle tracking PALM	Combines PALM with live-cell single fluorescent particle tracking.		
Stimulated emission depletion (STED)	Reduces the excitation volume below that dictated by the diffraction limit by coaligning one beam of light capable of fluorophore excitation with another that induces de-excitation by stimulated emission.	~16 nm	~10 min
Stochastic optical reconstruction microscopy (STORM)	Small sub-populations of photoswitchable fluorophores are turned on and off using light of different colours, permitting the localisation of single molecules. Repeated activation cycles produce a composite image of the entire sample.	<20 nm	~mins

Photoactivatable or "optical highlighter" fluorescent proteins (FPs) have emerged as powerful new tools for cellular imaging [38–47]. The fluorescent properties of these proteins can be altered upon illumination at specific wavelengths. They either switch between a fluorescent and non-fluorescent state (photoswitching) [48–54], or they change their fluorescence emission from one wavelength to another (photoconversion) [39,44,55,56]. The controlled photoconversion/switching of these proteins provides unique opportunities to mark and track selected molecules in cells in space and time [42,48,50,57–59]. High-density mapping of single-molecule motions can be obtained using photoactivated localisation microscopy (PALM) [60–62].

Another promising application of photoswitchable proteins is their use in super-resolution microscopy. This technique relies on the stochastic photoactivation and localisation of single molecules, in which a fluorescence image is constructed from high-accuracy localisation of individual fluorescent molecules that are switched on and off optically [63–68]. Microscope techniques that are based on this principle are called RESOLFT (reversible saturable optical fluorescence transitions) microscopy. RESOLFT microscopy concepts are photoactivated localisation microscopy (PALM) [63,65], fluorescence photoactivation localisation microscopy (FPALM) [69], stochastic optical reconstruction microscopy (STORM) [64,69–72], and PALM with independently running acquisition

(PALMIRA) [73,74] (Table 2). Image resolution well below the Abbe diffraction limit is achieved. Labelled proteins can be localised with a precision down to about 2–10 nm. Stunning images have been obtained based on photoactivatable FPs [63,75] (Figure 1D). Recently, super-resolution microscopy has also been extended to dual-colour imaging [76]. Another recently developed super-resolution technique is STED (stimulated emission depletion) [72,77]. In a STED microscope, the focal spot of excitation light is overlapped with a doughnut-shaped spot of light of lower photon energy, quenching excited molecules in the excitation spot periphery by stimulated emission. A resolution of 15 to 70 nm has been realised to map, for example, the nanoscale distribution of proteins inside cells [78], on the plasma membrane [79], and the movement of synaptic vesicles inside the axons of cultured cells [80].

2.1.3. Electron Microscopy

Microscopes consist of an illumination source, a condenser lens to converge the beam on the sample, an objective lens to magnify the image, and a projector lens to project the image onto an image plane, which can then be photographed or stored. In electron microscopes, the wave nature of the electron is used to obtain an image. There are two important forms of electron microscopy: scanning electron microscopy (SEM) and transmission electron microscopy (TEM). Both use electrons as the source for sample illumination. The lenses used in electron microscopes are electromagnetic lenses. For high-resolution surface investigations, two commonly used techniques are scanning electron microscopy (SEM) (Figure 1C) and AFM. The operation of the SEM consists of applying a voltage between a conductive sample and filament, resulting in electron emission from the filament to the sample. This occurs in a vacuum environment. The electrons are guided to the sample by a series of electromagnetic lenses in the electron column. The resolution and depth of field of the image are determined by the beam current and the final spot size. The electrons interact with the sample within a few nanometers to several microns of the surface, depending on the beam parameters and sample type. Along with the secondary electron emission (which is used to form a morphological image of the surface in the SEM), several other signals are emitted as a result of the electron beam impinging on the surface. Each of these signals carries information about the sample that provides clues to its composition. Two of the most commonly used signals for investigating composition are X-rays and backscattered electrons. X-ray signals are commonly used to provide elemental analysis. The percentage of beam electrons that become backscattered electrons has been found to be dependent on the atomic number of the material, which makes it a useful signal for analysing the material composition.

Since electron microscopy is conducted in a vacuum environment, it is at a disadvantage for the study of hydrated samples. To image poorly-conductive surfaces without sample charging may require conductive coatings or staining (which may alter or obscure the features of interest), or it may require low-voltage operation or an environmental chamber, which may sacrifice resolution. Recently, an electron microscopy technique was described for imaging whole cells in liquid that offers nanometer spatial resolution and a high imaging speed using a scanning transmission electron microscope (STEM) [81,82]. The cells were placed in buffer solution in a microfluidic device with electron-transparent windows inside of the vacuum of the electron microscope.

In TEM, the transmitted electrons are used to create an image of the sample. Scattering occurs when the electron beam interacts with matter. Scattering can be elastic (no energy change) or inelastic (energy change). Elastic scattering can be coherent and incoherent (with and without phase relationship). TEMs with resolving powers in the vicinity of 1 Å are now common. A relatively recent electron microscopy technique that can be used to study cells at the nanoscale is electron tomography. Electron tomography (ET) is the most widely applicable method for obtaining three-dimensional information by electron microscopy [83–85]. A tomogram is a three-dimensional volume computed from a series of projection images that are recorded as the object in question is tilted at different orientations. ET has the potential to fill the gap between global cellular localisation and the detailed three-dimensional molecular structure, because it can reveal the localisation within the cellular context at true molecular resolution and the shapes and three-dimensional architecture of large molecular machines. It can also reveal the interaction of individual proteins and protein complexes with other cellular components, such as DNA and membranes. A recent development is cryo-electron tomography (cryo-ET), which allows the visualisation of cellular structures under close-to-life conditions [86–88] (see Figure 1B as an example). Rapid freezing followed by the investigation of the frozen-hydrated samples avoids artifacts notorious to chemical fixation and dehydration procedures. Furthermore, the biological material is observed directly, without heavy metal staining, avoiding problems in interpretation caused by unpredictable accumulation of staining material. Consequently, cryo-ET of whole cells has the advantage that the supramolecular architecture can be studied in unperturbed cellular environments.

The ultrastructure of yeast cells (the model yeasts *S. cerevisiae* and *Sc. pombe*) was first studied by TEM using thin sections in 1957 [14], and the freeze-etching replica method was introduced in 1969 to obtain the fine structure of yeast cells [89]. During the next 50 years, techniques for the analysis of the ultrastructure of yeasts advanced greatly [90]. Initially, yeast cells were fixed solely with potassium permanganate ($KMnO_4$), and not by the widely used osmium tetroxide (OsO_4), since the thick cell wall is a barrier for the penetration of OsO_4. Finer EM images were obtained by using a double fixation with glutaraldehyde (GA) and $KMnO_4$ [91]. Important landmark studies have used conventional chemical fixation using GA and OsO_4—after enzymatic removal of the cell wall—to describe the cellular features of *S. cerevisiae* and to compare ultrastructural defects that result from mutations in key genes [92–96]. Next, methods using cryo-immobilisation followed by freeze substitution have been developed to provide excellent preservation of intact yeast cells [97–99]. These approaches involve rapid freezing of the sample with subsequent substitution treatment to replace frozen water in the sample with an organic solvent and fixatives [100]. Currently, high pressure freezing followed by freeze substitution (HPF/FS) is the method of choice for preparing cells for ET. Yeast prepared with these methods are used in 3D electron tomography studies for which sampling of the cell is performed at unprecedented resolution [88,101] (Figure 1B).

2.2. Force Microscopy

AFM techniques have turned out to be a suitable and versatile tool for single-molecule interactions (Table 3) and for probing the physical properties of microbial cell surfaces [102]. Especially, it has been used to study yeast surfaces: to determine nanomechanical properties of the cell wall, map cell wall proteins (Figure 2A), molecular recognition forces (receptor–ligand interaction), and characterise biomolecules by single-molecule unfolding (Table 4). For these types of analyses, the force sensing capabilities of the AFM are used. AFM-based force spectroscopy exerts pulling forces on a single attached molecule by retraction of the tip in the *z* direction (perpendicular to the *x*–*y* scanning plane). Cantilever bending is detected by the deflection of a laser beam onto a position-sensitive detector, such as a quadrant photodiode. A piezoelectric actuator stage is used to control the positioning of the sample relative to the tip. AFM-based force spectroscopy is also used to study single cell interactions (cell–cell and cell–substrate adhesion).

Table 3. Examples of yeast molecule interaction studies using single-molecule force spectroscopy (SMFS).

Cell Type	Interacting Molecule 1	Interacting Molecule 2	Rupture Force (pN)	Refs
C. albicans	Als5p	Fibronectin	2800 ± 600	[103]
S. cerevisiae	Prion protein Sup35 hexapeptide	Prion protein Sup35 hexapeptide	—	[104]
	Prion protein Sup35 hexapeptide antiparallel hairpin structure	Prion protein Sup35 hexapeptide antiparallel hairpin structure	32–134	[105]
	Nucleoporin	Nucleoporin	—	[106]
	Nucleoporin	Importin	—	
	Flo1p	Flo1p	300 (100–600)	[107]

Table 4. Examples of yeast receptor–ligand interaction studies using single-molecule force spectroscopy (SMFS) on cell surfaces.

Cell Type	Cell Receptor	Ligand	Rupture Force (pN)	Refs
	Cell surface β-mannan	Anti-β-1,2-mannoside antibodies	41 ± 14	[108]
	Cell surface β-glucans	Anti-β-1,3-glucan antibodies	38 ± 10	[108]
C. albicans	Cell wall chitin	WGA [1] lectin	65 ± 19	[108]
	Cell wall	*Streptococcus mutans* exoenzyme glycosyltransferase B	1000–2000	[109]
	Epa6p	Hydrophobic surface	-	[110]
C. glabrata	Cell surface β-mannan	Anti-β-1,2-mannoside antibodies	54 ± 9	[108]
	Cell surface β-glucans	Anti-β-1,3-glucan antibodies	41 ± 8	[108]
	Cell wall chitin	WGA [1] lectin	41 ± 8	[108]
	Cell surface α-mannan	Con A [2] lectin	75-200	[111]
	Cell surface α-mannan	Con A [2] lectin	92 ± 35	[108]
	Cell surface β-glucans	Anti-β-1,3-glucan antibodies	42 ± 7	[108]
S. cerevisiae	Cell wall chitin	WGA [1] lectin	54 ± 19	[108]
	Wsc1p-His-tagged	NTA-Ni^{2+}	-	[112–114]
	HA [3]-tagged Ccw12p	Anti-HA antibody	69.3 ± 31.4	[115]
	Ste2p	α-factor	250	[116]
S. pastorianus	Flo protein	Glucose	121 ± 53	[111]
	Flo protein	Con A [2]	117 ± 41	[111]

[1] WGA: wheat germ agglutinin; [2] Con A: concanavalin A; [3] HA: human influenza hemagglutinin.

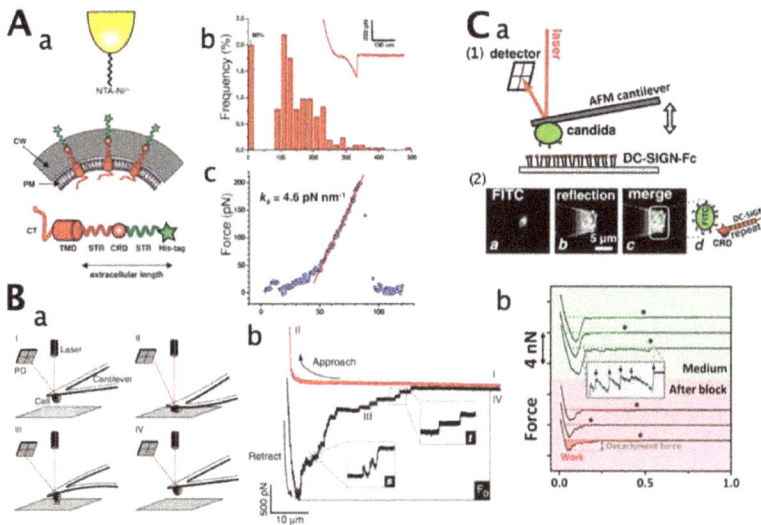

Figure 2. (**A**) (**a**) His-tagged modified Wsc1 membrane sensors were detected using AFM tips functionalised with Ni^{2+}-nitrilotriacetic acid (NTA) groups on live cells. The drawing shows a His-tagged elongated Wsc1 sensor with the cytoplasmic tail (CT), the transmembrane domain (TMD), the cysteine-rich domain (CRD), the serine/threonine-rich (STR) region, and the terminal His tag (in green) (CW = cell wall, PM = plasma membrane); (**b**) Adhesion force histograms and representative force curves recorded with a Ni^{2+}-NTA tip for *S. cerevisiae* cells expressing His-tagged elongated Wsc1 sensors; (**c**) Representative force extension curves obtained upon stretching a single Wsc1p. The curve displays a linear region, where force is directly proportional to extension. Reprinted with permission from [113]. (**B**) Experimental single-cell force spectroscopy (SCFS) setup. (**a**) A cell is attached to a coated cantilever. To measure the force acting on the cantilever, cantilever deflection is determined using a laser beam reflected by the cantilever onto a photodiode (PD). The cantilever-bound cell is lowered toward the substrate (I) until a preset force is reached (II). After a given contact time, the cantilever is retracted from the substrate (III) until cell and substrate are completely separated (IV);

(**b**) Force–distance (*F–D*) curve showing steps (I), (II), (III), and (IV), corresponding to those outlined in (a). Several unbinding events can be observed (*s*, force steps; *t*, unbinding of membrane tethers; F_D, maximal detachment force). Reprinted with permission from ref. [117]. (**C**) (**a**) (1) Interaction forces between the *C. albicans* cell and a dendritic cell-specific intercellular cell adhesion molecule-3 (ICAM-3)-grabbing non-integrin (DC-SIGN)-Fc-coated substrate, (2) a single fluorescein isothiocyanate (FITC)-labelled *C. albicans* cell immobilised on the apex of a tipless cantilever visualised by confocal microscopy. Single channels (*a–b*) and an overlay (*c–d*) show the FITC-labelled *C. albicans*; (**b**) Probing specific DC-SIGN–*C. albicans* interactions with atomic force microscope dynamic force spectroscopy. Examples of *F–D* curves of the interaction of DC-SIGN with *C. albicans*; single bond ruptures are visible as discrete steps (arrows in inset). The area enclosed by the curve and the zero-force line (no contact regime; dotted line) is a read-out for the adhesion between the cell and the substrate; the work needed to detach *C. albicans* from the DC-SIGN-Fc-coating. Next, three examples of *F–D* curves are shown after an in situ block with soluble *C. albicans* (CA)-mannan. The work and detachment force (indicated maximum force F_{max}) are smaller than before this block. The asterisk (*) indicates the distance at which the final bond detaches. The arrows in the inset indicate discrete rupture steps. Reprinted with permission from ref. [118].

Single-cell force spectroscopy (SCFS) assays on living cells have been applied to measure the strength of cell adhesion down to the contribution of single molecules [119–121] (Figure 2B,C). AFM-based SCFS is currently the most versatile method for the study of adhesive interactions of cells with other cells, proteins, and surfaces, since SCFS offers a large range of detectable forces (from 10 pN to 100 nN), and offers precise spatial (1 nm to 100 μm) and temporal (0.1 to >10 min) control over the adhesion experiment and experimental parameters [120]. A living cell can be attached to a tipless AFM cantilever and the interacting partner (molecule or cell) on a substrate-coated surface. Alternatively, the living cell can be fixed on a surface, and the tip functionalised with the interacting molecule. AFM force spectroscopy with a single cantilever-bound cell can be used to investigate cell–cell and cell–matrix interactions. The approach and withdrawal of this cell to and from its surface can be precisely controlled by parameters such as applied force, contact time, and pulling speed, benefiting from the AFM's high-force sensitivity and spatial resolution. The data collected in these experiments include information on repulsive forces before contact, cell deformability, maximum unbinding forces, individual unbinding events, and the total work required to remove a cell from the surface (Table 5, Figure 2B,C). Force spectroscopy can identify cell subpopulations and characterise the regulation of cell adhesion events with single-molecule resolution [122].

Table 5. Examples of yeast using single-cell (SCFS) force spectroscopy studies.

Cell Type	Interaction Partner	Variables	Refs
C. albicans	Staphylococcus aureus	Deletion of *ALS3* (adhesion gene)	[123]
	Hydrophobic DDP [1] coated surface	Surface hydrophilicity, hydrophobicity, deletion of *HGC1*, compared to *S. cerevisiae*	[124]
	C. albicans hyphae	Deletion of *ALS3* and *ALS1* (adhesion genes)	[125]
	DC-SIGN [2]	Differences in the N-mannan structure of the cell wall	[118]
C. glabrata	Adhesin Epa6p	Surface hydrophilicity, hydrophobicity, expressed and deleted *EPA6*	[110]
S. cerevisiae	Abiotic surface	Surface hydrophilicity, hydrophobicity BSA coating, life cycle stage, glutaraldehyde-treated cells	[126]
	Silica surface	Different silica with defined roughness	[127]
	Methacrylate polymers surface	Polymer imprinted and non-imprinted surface	[128]
	Bare and polydopamine-coated glass	Polydopamine coating	[129]

[1] DDP: dodecyl phosphate; [2] DC-SIGN: dendritic cell-specific intercellular cell adhesion molecule-3 (ICAM-3)-grabbing non-integrin.

2.3. Nanomotion Analysis

New sensor technologies based on microcantilevers have recently been developed [130,131]. Nanomechanical oscillators are increasingly being used for the detection of very small masses [132] or for nanostress sensing in molecular biology [133,134]. Cantilever resonators have been shown to possess a mass resolution in the pico- to femtogram ranges in both air [135] and liquid [136,137]. Many of the available systems are limited by the need to perform the measurements in air or in a humid environment, and most rely on the detection of the replication of the cells on the surface of the sensor. Thanks to the many advantages they offer, microcantilevers have recently been explored as nanosensors for cell studies; they are highly sensitive, selective, label-free, real-time, and provide in situ detection capabilities [138]. Single cell detection and monitoring on the cantilever sensor has been reported for *S. cerevisiae* cells [139,140], *E. coli* and *Bacillus subtilis* [140,141], HeLa cells [142], mouse lymphoblasts [140], and human lung carcinoma and mouse lymphocytic leukemia cells [143]. Cell growth detection has been demonstrated by monitoring resonance frequency changes of cantilevers as the mass increases from immobilized *S. cerevisiae* and fungal *Aspergillus niger* spores on the surface of the cantilevers in humid air [144]. *S. cerevisiae* cells were deposited onto the cantilever surface, and its bending as a function of time corresponded to the yeast growth behaviour [138].

Recently, the metabolic state of living organisms that are immobilized on the cantilever surface could be detected by cantilever nanomotion analysis in physiological conditions [145–147]. In nanomotion analysis mode, the sample is directly deposited onto the cantilever, and the analysis is performed with the functionalised cantilever in liquid. This differs from nanomechanical resonators, where the liquid sample is flowed through a capillary in the cantilever (Figure 3A). If the sample is alive, its nanometric-scale motions are transmitted to the cantilever, causing it to oscillate. These oscillations are detected by monitoring the cantilever displacements with the traditional laser–photodiode system; a typical set-up is depicted in Figure 3Ba. The cantilever and the sample of interest are immersed in an analysis chamber equipped with an inlet and an outlet that permits measurement in liquids, and, importantly, the exchange of liquids during measurements. It has been observed that any type of organism induces oscillations of the cantilever that only last while the organism is alive [147]. Once an efficient killing agent is applied, the cantilever oscillations stop. The exact origin of these vibrations is still under investigation. In the case of motile organisms, such as mammalian cells or flagella-equipped bacteria such as *E. coli*, the answer is straightforward. However, in the case of immotile microorganisms such as yeast or *Staphylococcus aureus*, the explanation is more challenging. Probably, a direct momentum transfer between the sample's surface proteins that undergo conformational changes and the cantilever plays an important role [146].

Figure 3Bb shows a typical nanomotion experiment with *C. albicans*. The AFM cantilever was pre-treated with glutaraldehyde and incubated in a solution containing the cells. Some *Candida* cells attached onto its surface. The cantilever was eventually inserted into the growth medium-filled analysis chamber, and its oscillations were recorded. After the injection of a buffer solution containing 10 μg/mL of caspofungin (an antifungal drug to which *Candida* is sensitive) in the analysis chamber, the cantilever oscillations dramatically decreased. This drop became noticeable after only 10 min post-caspofungin exposure. Such an application can be very efficient (in a timeframe of minutes) for the detection of chemicals to which living organisms are sensitive, or for simple assessment of the presence of living organisms in extreme environments.

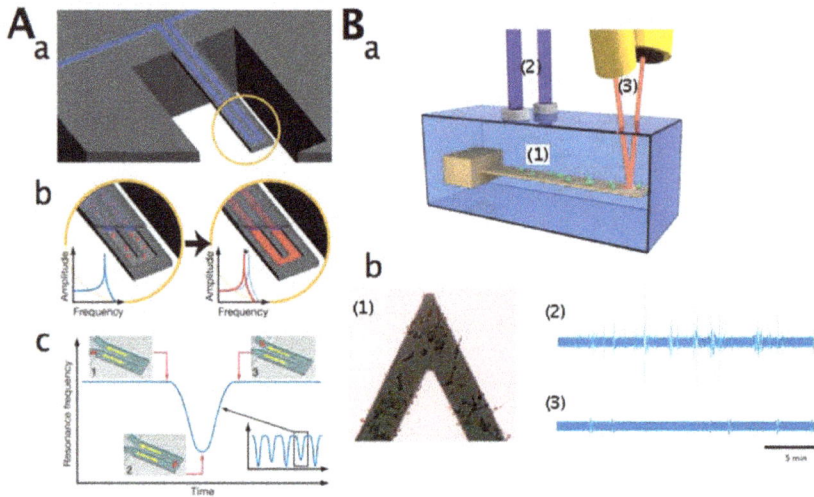

Figure 3. (**A**) Nanomechanical resonators enable the measurement of mass with extraordinary sensitivity. Illustration of two mass measurement modes enabled by a fluid-filled microcantilever. (a) A suspended microchannel translates mass changes into changes in resonance frequency. Fluid continuously flows through the channel and delivers biomolecules, cells, or synthetic particles. Sub-femtogram mass resolution is attained by shrinking the wall and fluid layer thickness to the micrometre scale and by packaging the cantilever under high vacuum; (b) While bound and unbound molecules both increase the mass of the channel, species that bind to the channel wall accumulate inside the device, and, as a result, their number can greatly exceed the number of free molecules in solution. This enables specific detection by way of immobilised receptors; (c) In another measurement mode, particles flow through the cantilever without binding to the surface, and the observed signal depends on the position of particles along the channel (insets 1–3). The exact mass excess of a particle can be quantified by the peak frequency shift induced at the apex. Reprinted with permission from [141]. (**B**) (a) Typical setup for the detection of the nanomotion of living organisms suspended in liquid medium: (1) analysis chamber with microbial cells (green) attached to the cantilever, (2) inlet and outlet of the fluid chamber, (3) laser and photodetector; (b) (1) *C. albicans* deposited onto a cantilever. Courtesy of Dr. Sandor Kasas, Ecole Polytechnique Fédéral de Lausanne, Switzerland, (2) oscillations of the cantilever in nourishing medium-filled analysis chamber, (3) oscillations of the cantilever after the replacement of the growth medium with caspofungin (antifungal agent)-containing buffer. The amplitude of cantilever oscillations is in the range of 1–8 nm (unpublished data).

3. Yeast Cell Patterning and Manipulation

3.1. Yeast Cell Patterning

Manipulating the physical location of cells is useful both to organize cells in vitro and to separate cells during screening and analysis [148,149]. The quest to manipulate cells on length scales commensurate with their size has led to a host of technologies exploiting chemical, mechanical, optical, electrical, and other phenomena. The major cell-patterning methods include patterning on adhesive micropatterns, mechanical cell patterning, and robotic cell patterning [150]. Cell-adherence methods have been especially developed for the adhesion of mammalian cells, but have also been developed for yeast cell patterning (Table 6). A variety of different patterning techniques have been developed to present adhesive ligands at a range of scales to investigate biological events, pushing the envelope on the minimum feature down to the nanometer scale [151–156].

Table 6. Applications of yeast cell patterning.

Yeast Type	Cell Patterning Method	Issue Addressed	Refs
S. cerevisiae	Patterning on adhesive micropatterns	Microcontact printing of concanavalin A	[157]
	Mechanical cell patterning in microfluidic microchambers	Monitoring dynamics of single-cell gene expression	[158]
	Robotic cell printing	Systematic profiling of cellular phenotypes	[159]
	Mechanical cell patterning using trap barriers	Single cell gene expression analysis	[160]
	Robotic cell printing	Localisation of the yeast proteome during polarised growth	[161]
	Mechanical cell patterning in microfluidic microchambers	Quantitative analysis of the yeast pheromone signalling response	[162]
	Patterning on adhesive micropatterns	Microcontact printing of biotinylated bovine serum albumin	[163]
	Mechanical cell patterning using trap barriers	Whole lifespan microscopic observation	[164]
	Mechanical cell patterning in microfluidic single-cell microwells	Real-time cellular responses of the mating MAPK pathway	[165]
	Mechanical and chemical patterning in microchambers	Molecular phenotyping of aging in single cells	[166]
	Mechanical cell patterning using trap barriers	Single cell analysis of yeast replicative aging	[167]
	Mechanical cell patterning in elongated cavities	Monitoring the dynamics of cell division	[168]
	Mechanical cell patterning using trap barriers	Studying ageing and dynamic single-cell responses	[169]
	Patterning in microcavity array by negative pressure, and embedded in agarose gel layer	Long-term single cell growth observation	[170]
	Mechanical cell patterning using trap barriers	Automated measurements of single-cell aging	[171]
	Mechanical cell patterning using trap barriers	High-throughput analysis of yeast replicative aging	[172]
Sc. pombe	Mechanical patterning in culture microchambers	Mechanical mechanisms redirecting cell polarity and cell shape in fission yeast	[173]
	Mechanical patterning in single-cell microwells	Determination of the mechanical forces involved in cell growth	[174]
	Mechanical patterning in microchambers	Time-lapse fluorescence observation of the effect of a microtubule-inhibiting drug	[175]
	Mechanical trapping in single-cell cavities	Fission yeast synchronisation	[176]
	Mechanical barrier single-cell trapping	Lon-term observation using super-resolution fluorescence microscopy	[177]
	Mechanical patterning in chemostat microchambers	Long-term single-cell analysis	[178]
	Mechanical patterning in culture microchambers	Studies of cellular aging	[179]

MAPK: Mitogen-activated protein kinases.

Microcontact printing has become the most popular technique [180]. A polydimethylsiloxane (PDMS) stamp with desired microfeatures is fabricated using soft lithography methods, and is used to print adhesive biomolecules onto the culture substrate [157,181]. For yeast cell adhesion, the lectin concanavalin A (which binds to cell wall mannose and glucose aminoglycans) can be used as an adhesive molecule. *S. cerevisiae* was also immobilised on cholesterol-modified microcontact-printed spots [163]. Despite its popularity, microcontact printing has several drawbacks for cell biology labs, such as the requirement of a clean room to microfabricate the stamp, and variations in the quality of the protein transfer [182].

In mechanical cell patterning, mechanical barriers capture the cells at specified spots. Cells can be trapped in microchambers (Figure 4A), microwells (Figure 4C), or by cell trap barriers (Figure 4D) (Table 6). Various microfabrication techniques have been used to fabricate microwell substrates for cell cultivation [150]. The microwell can have a diameter from several hundred micrometers up to the dimensions of a single cell [174]. Single-cell microwell arrays allow large numbers of cells to be stimulated and analysed (usually by fluorescence microscopy) in a massively parallel fashion [165,183]. Single-cell analysis has increasingly been recognised as the key technology for the elucidation of cellular functions, which are not accessible from bulk measurements on the population level [184,185]. Yeast cells have been trapped in microfluidic microchambers by using inlet and outlet valves [162,186] (Figure 4A). Culture chambers that are open on both sides [173,179,187] or on one side [176] have been constructed. These chambers fit single-cell dimensions and confine the cells. These culture chambers are suitable for non-adherent cells, such as yeast and bacteria [188].

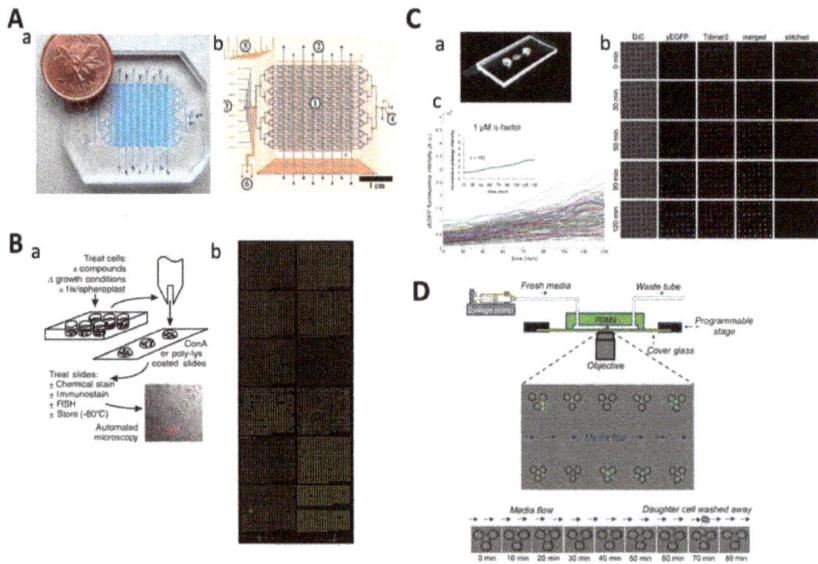

Figure 4. (**A**) Cell trapping in microfluidic chambers. (**a**) Image of the microfluidic device; (**b**) Working area of microfluidic device showing (1) array of 128 imaging chambers, (2) column inlets for loading different strains, (3) eight chemical inlets controlled by independent valves, (4) outlet ports, (5) fluidic multiplexer to deliver reagents to specified rows, (6) integrated peristaltic pump for on-chip formulation of stock reagents. Reprinted with permission from [162]. (**B**) (**a**) Spotted cell microarrays using contact cell printing. Cell chips are constructed using slotted steel pins to print cells robotically from multi-well plates onto glass slides; (**b**) Wide-field light scattering image of a cell microarray containing around 4800 viable haploid yeast deletion strains. From ref. [159]. (**C**) Single-yeast cell microwell array in a microfluidic chip. (**a**) An overview of the cell chip. The cell chip has one simple straight microfluidic channel and two punched reservoirs; (**b**) Representative microscopic images for 0, 30, 60, 90, and 120 min time points in the case of α-factor treatment (DIC: bright field, yEGFP: green fluorescent, and Tdimer2: red fluorescent images). The merged and stitched images show diverse colours from a mixture of green and red fluorescence; (**c**) Typical time-course measurements of mating responses of individual cells. The inset shows the normalised time-course average of yEGFP fluorescence intensity. Reprinted with permission from [165]. (**D**) A microfluidics platform that facilitates simultaneous lifespan and gene expression measurements of aging yeast cells. Schematics of the experimental setup (*upper panel*). The growth of a single cell that is trapped in a replicator as a function of time is shown (lower panel). From ref. [171].

Mechanical cell trap barriers have also been used to capture cells from suspensions in fluidic devices [158,189,190]. Fluid flow pushes the cells into the traps, and, therefore, these cell traps are also designated as hydrodynamic cell traps [191]. Barriers have been designed with a small fluidic leak that allowed single-cell trapping [160,177,192] (Figure 4D).

To create cellular microarrays, cells can be spotted or "printed" using a fluid-dispensing device ("cell printer") [150]. It is essential to obtain a highly reproducible number of living cells per spot and an optimised printing process that is qualified for the reproducible production of microarrays with cells that keep their vitality and function for analysis. Spot formation techniques are categorised as "contact printing" and "non-contact" printing [193,194]. Robotic yeast cell contact printing was initially used to print cells on an agar growth medium by using fluid-dispensing devices or pads [195], or cells were grown in multiwell culture plates and printed on a glass slide for high-throughput imaging [159] (Figure 3B), or only short-time analyses on living cells were performed. More often, non-contact-based devices are used to produce cellular arrays, such as modified inject printers or piezo-driven tips [196–199]. In non-contact printing techniques, the liquid metering is not determined by the complex interplay of the pin, the liquid, and substrate, but is separated from the substrate, because no contact between the printing tool and the substrate occurs. The fluid is ejected as a flying droplet or jet towards the surface from a certain distance, which makes metering more precise. One concept of non-contact printing is based on syringe–solenoid-driven printers, where a reservoir and a high-speed microsolenoid valve are connected to a high-resolution syringe (e.g., the M2-Automation, synQUAD, or Genomic Solutions system). Further non-contact microarrayers are piezoelectrically driven, where a technology similar to the one used in an ink-jet printer is used (e.g., M2-Automation, MicroDrop, PerkinElmer, Scienion, GeSim) [200,201]. A piezo-actuator is fixed at the top of the dispenser tip. The squeezing of the tip forced by the piezo-actuation induces droplet ejection out of the capillary. The fast response time of the piezoelectric crystal permits fast dispensing rates (kHz range), and the small deflection of the crystal generates droplets from tens of picoliters to a few nanoliters.

3.2. Direct Contact Cell Manipulation

3.2.1. AFM-Based Cell Manipulation

The desire to actively deliver precise amounts of biomolecules through nanosized probes initiated the development of novel microfluidic probes. Microfabrication processes have been introduced for the production of AFM cantilevers with embedded microchannels [202–205]. Microchannel cantilevers were connected to a pressure controller for active liquid handling in fluidic force microscopy (FluidFM) [206,207]. The ability to apply a pressure allows for negative pressure experiments involving suction for applications such as cell adhesion, or positive pressure experiments resulting in cell deposition on a specified spot or in controlled dispensing for applications such as the accurate delivery of bioactive compounds to a single targeted cell in physiological medium or even cell injection. FluidFM was used for the spatial manipulation of single *S. cerevisiae* cells [208]. Therefore, the hollow cantilever was positioned over a yeast cell and approached in AFM contact mode. An underpressure of ~50 mbar was applied to suck the cell against the channel aperture. After displacement, the cell was deposited onto the substrate with an AFM approach in contact mode, and the cell was released by applying a short overpressure pulse while retracting the probe. The underpressure single-cell immobilisation of cells on the cantilever also allows accelerating the pace of SCFS, since the conventional cell trapping cantilever chemistry can be avoided [124]. Single-cell *C. albicans* adhesion forces to a hydrophobic (dodecyl phosphate coated) surface were compared to adhesion to a hydrophilic (hydroxyl-dodecyl phosphate coated) surface, the *C. albicans* mutant Δ*hgc1* (which reduces the cell surface hydrophobicity), and to *S. cerevisiae* adhesion to the hydrophobic and hydrophilic substrate (Table 5). Force adherence measurements of *S. cerevisiae* cells on bare glass and polydopamine-coated glass substrates have been performed using a microfabricated hollow cantilever made entirely from SU-8 [129] (Table 5). Highly flexible SU-8 cantilevers with integrated microchannels have been fabricated for both additive and subtractive patterning of *S. cerevisiae* cells [209].

3.2.2. Micropipette Manipulation of Single Yeast Cells

The oldest and most commonly used approach for single-cell manipulation uses glass capillary micropipettes [210]. A negative pressure applied to growth media-filled capillary immersed in a cell culture dish controls the aspiration of a desired cell. A positive pressure dispenses the cell. Motion stages with multiple degrees-of-freedom were used to manually manipulate the micropipette and accurately control its tip position to perform either micromanipulation or microinjection [211]. Micromanipulators enable the controlled separation of selected living cells from suspension and even allow for isolation of prokaryotic cells [212]. They can also be used in adhesion studies, such as the interaction of a single *C. albicans* cell that is sucked to a micropipette with a diameter that is smaller than the cell, with a salivary pellicle-coated bead that is manipulated with a second micropipette [213].

Single cell manipulation systems that are based on capillaries are commercially available; for example: TransferMan (Eppendorf, Hamburg, Germany), PicoPipet (Bulldog Bio, Portsmouth, NH, USA), Stoelting Micromanipulators (Wood Dale, IL, USA), and miBot™ manipulator (Imina Technologies, Lausanne, Switzerland). These manipulation systems are manual, although the miBot micromanipulator is a mobile micro-robot that moves directly over the surface of the microscope base, has a nm spatial resolution, and can be remotely controlled. Micropipette cell manipulation systems that allow automatic selection and placement of a single yeast cell using vision-based feedback control have been developed [211]. A robotic micromanipulation system based on a general-purpose micromanipulator and a traditional glass micropipette was developed for pick-and-place positioning of single cells [214]. By integrating computer vision and motion control algorithms, the system visually tracks a cell in real time and controls multiple positioning devices simultaneously to accurately pick up a single cell, transfer it to a desired substrate, and deposit it at a specified location. A computer-controlled micropipette installed on an inverted fluorescence microscope was used to automatically recognise by computer vision, and both fluorescently labelled and unlabelled live cells in a Petri dish were picked up [215]. A recent developed computer vision-based automated single-cell isolation system allowed the isolation of single live cells from a very dense culture without immobilising cells on a surface [216].

Microchanneled AFM micropipettes have also been developed and used for cell adhesion and spatial cell manipulation applications (see previous section "AFM-based cell manipulation"). These AFM micropipettes are also designated as versatile nanodispensing (NADIS) systems [217]. Compared to conventional glass pipettes, this tool is particularly suitable when using substances of high cost or limited amounts, because significantly less volume is required for an experiment [218,219]. Another advantage over glass pipettes is the precise control wielded in the manipulation of sensitive targets, due to concurrent measurements of cantilever deflections without significant target damage [206]. Targets—such as functionalised surfaces or surface immobilised cells—can be precisely and gently manipulated physically, biologically, and chemically [129,207,208,220].

3.3. Non-Contact Cell Manipulation

3.3.1. Optical Manipulation of Single Yeast Cells

In the last decade, optical manipulation has evolved from a field of interest for physicists to a versatile tool widely used within life sciences [221]. Optical trapping and manipulation is a spin-off from research where lasers were used to study the effect of linear and angular momentum of light on small neutral particles. Arthur Ashkin first demonstrated that radiation pressure from a focused laser beam significantly affected the dynamics of micrometer-sized transparent and neutral particles, and two basic light-pressure forces were discovered: a scattering force in the direction of the incident light beam, and a gradient force in the direction of the intensity gradient of the beam [222]. The scattering component of the force works as a photonic "fire hose" pushing the particle in the direction of light propagation. The gradient force can be explained by a dipole in an inhomogeneous electric field that experiences a force in the direction of the intensity field gradient of the laser beam [223]. Using these forces, small particles (such as cells) can be accelerated, decelerated, and trapped in three dimensions.

Optical tweezers use light to levitate a particle (cell) of distinct refractive index [224]. The trapped cell is suspended at the waist of the focused (typically infrared) laser beam. The displacement of the cell from the focal center results in a proportional restoring force, and can be measured by interferometry or back-focal plane detection. Optical tweezers use a high gradient of optical pressure to guide cells by focusing a laser beam through a high numerical aperture (N.A.) lens on the cells. High optical intensity of about 10^{10} mW/cm^2 may cause damage on the cells, and are not suitable for long-term cell manipulation [225]. Micro-meter-sized homogeneous particles (or cells) can be trapped with forces ranging from a few pN to several tens of pN, depending on the optical properties of the particles and of the medium [226]. It is possible to track the position of the trapped particle with sub-nanometer accuracy at high (several MHz) repetition rates [227]. Due to the rapid advances in laser technology, optical manipulation setups have been developed that have become relatively uncomplicated. Optical manipulation is easily integrable with various microscopy setups, including confocal, super-resolution, or multiphoton microscopes. It allows for high spatial and temporal resolution, and interaction forces can be minimised.

Optical tweezers have been used to manipulate yeast cells, such as in cell trapping, cell positioning, and cell sorting (Table 7, Figure 5). Optical trapping was used to isolate single yeast cells from a mixture of two strains that were distinguishable in fluorescence microscopy [228]. An optical tweezer was used for the rapid separation and immobilisation of a single yeast cell by concomitant laser manipulation and locally thermosensitive hydrogelation [229]. Optical tweezers can be used to trap single yeast cells for further analysis, such as Raman microspectroscopy [230], time-lapse fluorescence microscopy to determine single-cell internal pH [231] (Figure 5C), or to study single cell dynamics by monitoring GFP-tagged proteins [232] (Figure 5D). Yeast cells are conducive to direct optical tweezing and can be used for single-cell force studies [233]. The effect of various factors (such as the ionic strength and the nature of the counter-ion in the solutions) on the adhesion and detachment force of yeast cells on glass was assessed [234]. Compared to AFM, magnetic tweezers, and more conventional ways of studying cell adhesion (such as shear-flow cells), optical tweezers present several advantages: direct measurements in physiological conditions, clear criterion to evaluate the proportion of adhering cells, and ease of examining the heterogeneity of cell behaviours in the population. However, the optical tweezer method is limited to low adherence forces (~1 to 100 pN) owing to the low refractive index of cells, and is sensitive to the cell optical heterogeneity.

Optical gradient forces generated by fast steerable optical tweezers are highly effective for sorting small populations of cells in a lab-on-a-chip environment (Figure 5). Reliable sorting of yeast cells in a microfluidic chamber by both morphological criteria and by fluorescence emission was demonstrated [235]. More than 200 yeast cells could be contact-free immobilised into a high-density array of optical traps in a microfluidic chip [236] (Figure 5B). The cell array could be moved to specific locations on the chip, enabling the controlled exposure of cells to reagents and the analysis of the responses of individual cells in a highly parallel format using fluorescence microscopy. Additionally, single cells were sorted within the microfluidic device using an additional steerable optical trap. Optical tweezers were used to spatially and temporally control pathogenic *C. albicans* and *Aspergillus fumigatus* and place them in proximity to host cells, which were subsequently phagocytosed [237,238].

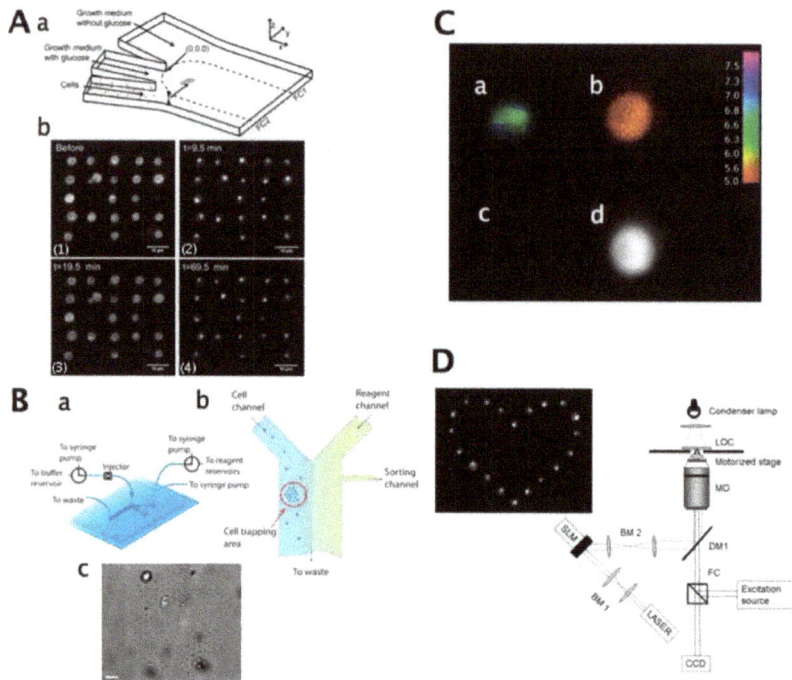

Figure 5. (**A**) (**a**) Schematic of the microfluidic device for subjecting single cells to environmental changes. The cells are collected in the flow of cells from the lower channel using optical tweezers and positioned within the measurement region. By changing the relative flow rates at the inlets (from FC1 to FC2 and back), the environment around the cells can be changed reversibly; (**b**) Images showing yeast cells expressing Msn2-GFP: (1) to (4) show the cellular response (i.e., the shuttling of Msn2-GFP proteins in and out of the nucleus) when four cycles of changing the medium back and forth between 4% and 0% glucose was performed in the microfluidic chip. Reprinted with permission from ref. [239]. (**B**) (**a**) Schematic description of the microfluidic chip; and (**b**) the fluidic circuit; (**c**) Transmission micrograph of more than 200 optically trapped yeast cells; scale bar: 30 µm. Reprinted with permission from ref. [236]. (**C**) Optically trapped single *S. cerevisiae* cell. All four images are of the same cell. (**a,b**) show the pH distribution at t = 0 min (30 °C) and t = 12 min (70 °C); (**c,d**) are propidium iodide images at t = 0 min (30 °C) and t = 12 min (70 °C). The colour bar represents pH values. Reprinted with permission from ref. [231]. (**D**) Typical optical tweezer setup. The trapping laser light is guided onto the spatial light modulator (SLM) via a beam expander (BM1). The laser beam with the imposed phase pattern then passes a second beam expander (BM2) to be imaged onto the back focal plane of the microscope objective (MO). The schematic figure of the setup also shows the lab-on-a-chip (LOC), a fluorescent excitation light source, a filter cube (FC), a dichroic mirror (DM1), a motorised microscope stage, and a condenser lamp. The inset image shows holographically-trapped *S. cerevisiae* cells that were stressed with sorbitol to induce localisation of Hog1-GFP to the cell nuclei. Reprinted with permission from ref. [221].

Table 7. Examples of yeast cell manipulation using optical manipulation.

Yeast Type	Issue Addressed	Refs
C. albicans	Control and manipulation of pathogenic yeast for live cell imaging and interaction with host cells	[237,238]
Hanseniaspora uvarum and *S. cerevisiae*	Confinement of an individual *H. uvarum* cell by *S. cerevisiae* cells increases the average generation time	[240]
S. bayanus	Study of growth pattern of cells under line optical tweezers generated by time-shared multiple optical traps	[241]
S. cerevisiae	On-chip single-cell separation and immobilisation using optical manipulation and thermosensitive hydrogel	[229]
	Real-time detection of hyperosmotic stress response in optically trapped single yeast cells using Raman microspectroscopy	[230]
	Optical manipulation of cells to microscopically observe environmentally-induced size modulations and spatial localisation of GFP-tagged proteins to elucidate various signalling pathways	[232]
	Optical trapping and surgery of living cells using two operational modes of a single laser	[242]
	Selection and positioning of single cells combined with microscopy analysis in a microfluidic channel; cycling of GFP-tagged Mig1p and Msn1p between the cytosol and nucleus	[239]
	Optical trapping and fluorescence microscopy investigation of the internal pH response and membrane integrity with increasing temperature	[232]
	Automated transportation of single cells	[243]
	Development of a microfluidic array cytometer based on refractive optical tweezers for parallel trapping, imaging, and sorting of individual cells	[236]
	Microfluidic sorting of arbitrary cells with dynamic optical tweezers	[235]
	Development of graded-index optical fibre tweezers with long manipulation length	[244]
	Position yeast cells in a microfluidic chamber to study glycolytic oscillations	[245,246]
	Tomographic phase microscopy with live cell rotation using holographic optical tweezers	[247]
	Development of a photonic crystal optical tweezer to trap an array of yeast cells	[248]
Sc. pombe	Displacement of the lipid granules	[249]
	Displacement of the nucleus	[250,251]
	Laser ablation of microtubules in vivo	[226]
	In vivo anomalous diffusion and weak ergodicity breaking of lipid granules	[252]
	Quantitative determination of optical trapping strength and viscoelastic moduli inside living cells	[253]

Optical tweezers can also be used to investigate the complex system of mechanical interactions taking place inside a living cell [226,249,250]. The viscoelastic properties of living *Sc. pombe* were investigated by studying the diffusion of lipid granules naturally occurring in the cytoplasm [251,252]. Optical manipulation techniques, such as optical tweezing, mechanical stress probing, or nano-ablation allow handling of probes and sub-cellular elements (such as organelles and individual molecules) with nanometric and millisecond resolution [254]. A near-infrared optical tweezer was used for yeast cell manipulation and micro-ablation [255]. This micro-nanosurgery system is based on a pulsed ultraviolet laser that induces plasma formation for intracellular surgery in live culture cells with submicron precision. Optical tweezers allow force probing of organelles and single molecules in vivo [256,257]. PicoNewton forces—such as those involved in cell motility or intracellular activity—can be measured with femtoNewton sensitivity, while controlling the biochemical environment. A method to perform a correct force calibration inside a living yeast cell (*Sc. pombe*) was developed [257]. This method takes the viscoelastic properties of the cytoplasm into account, and relies on a combination of active and passive recordings of the motion of the cytoplasmic object of interest. Absolute values for the in vivo viscoelastic moduli of the cytoplasm as well as the force constant describing the optical trap were determined.

3.3.2. Electrical and Magnetic Manipulation of Yeast Cells

Instead of using surface chemistry to prevent or allow cells to attach to certain regions, electromagnetic forces can be used to control cell positioning [149] and adhesion [258,259]. Electrical fields are very suitable for cell and bioparticle manipulation, with the advantages of strong controllability, easy operation, high efficiency, and minimal damage to targets [260]. Electrokinetic motion of cells refers to the migration of electrically charged or uncharged particles in a liquid medium or suspension in the presence of an electric field [261]. Electrical forces for manipulating cells at the microscale include electrophoresis and dielectrophoresis (DEP) [148,149]. Electrophoretic forces arise from the interaction of a cell's charge and a uniform or non-uniform electric field, whereas dielectrophoresis refers to the motion of polarised (uncharged) particles in only a non-uniform electric field [262]. Based on the applied electric field, DEP can be broadly divided into AC (AC DEP, classical DEP), DC (DC DEP), insulator-based DEP (iDEP, DC-iDEP), combined AC/DC (AC-iDEP), and travelling wave DEP (twDEP). In AC DEP, an array of metal electrodes is embedded inside a microdevice (such as a microfluidic chip) to generate a spatially non-uniform electric field, and can be used to separate particles by changing the medium property and frequency of the applied electric field [263,264]. It can also eliminate any electrophoretic (EP) and electroosmotic (EO) effect [265]. In DC DEP, the spatially non-uniform electric field is created by specially-designed insulators, such as electrically non-conducting obstructions or hurdles in a microdevice, and electrodes that are positioned at the ends of the microfluidic channels [263]. The DC electric field results in an EO force and eliminates the need for an external pump, which is required in the case of AC DEP. The combination of AC/DC DEP can transport cells by using electrokinetic effects (such as electroosmosis), and AC DEP can be used to separate cells. In the travelling wave DEP, transport and separation of cells can be performed with the AC electric field [266]. In this case, the spatial nonuniformity of the phase of the electric field is used to transport the particle, and the nonuniformity in magnitude of the field is used to separate the particles.

DEP and electrophoretic forces have been used to create microsystems that separate cell mixtures into their component cell types or act as electrical "handles" to transport cells or place them at specific locations [267] (Figure 6C). DEP has also been applied for cell sorting [268], focussing, filtration [269], and assembly [262]. DEP has been used to characterise cells, for example, to monitor cell viability changes (including morphology and internal structure) and isolate viable cells with minimal or no damage [270,271]. Electrophoretic and/or electroosmotic pumping can also be used to control and drive cell transport in microfluidic chip channels [272]. DEP tweezers have been developed that allow the positioning of a single cell in three dimensions or transfer a single cell to any designated area [260]. A DEP tweezer consisting of a sharp-tip glass needle with a pair of electrodes and using a pDEP force could hold a single yeast cell at the end of the micromanipulator [267]. The design of the tweezers was not adequately optimised for one-by-one manipulation; therefore, a round-tip shape for the DEP-based tweezers that shifts the electric field to the centre of the tweezers' tip due to the smooth geometry at the tip is most suitable for single-cell manipulation [273].

DEP traps for single-cell patterning in physiological solutions have been developed [274] (Table 8). DEP manipulation and trapping of yeast cells has been included in microdevices such as microfluidic chips [272,275–277]. Live and dead yeast cell separation was achieved with the "headlands and bays" electrodes [278]. Live yeast cells are attracted to the regions of the maximum field, while the dead ones are repelled to the regions of the minimum field, resulting in the separation of live and dead yeast cells. The separation of live and dead yeast cells by DEP could be enhanced by using the cross-linking agent glutaraldehyde (since glutaraldehyde selectively cross-links nonviable cells to a much greater extent than viable cells due to the higher cell wall permeability of nonviable cells) [279]. Live and dead *S. cerevisiae* cells were sorted by using AC DEP [280], multifrequency DEP [281], or AC/DC DEP using a quadrupole electrode array [282] (Figure 6A). Yeast cells could be pre-concentrated by trapping using DEP, and separated depending on their vitality by using hydrodynamic, DC electrophoretic, and DC electroosmotic forces [271]. Live and dead yeast cells were characterised based on dielectric properties [283,284]. By combining DEP and image processing, dielectrophoretic spectra of cells can be acquired [285]. From this, the membrane properties of the cells can be obtained. DEP was used to control the rotation and vibration of patterned yeast cell clusters [286] (Figure 6B). This strategy is based on the cellular spin resonance mechanism, but it utilises coating agents to create consistent rotation and vibration of individual cells.

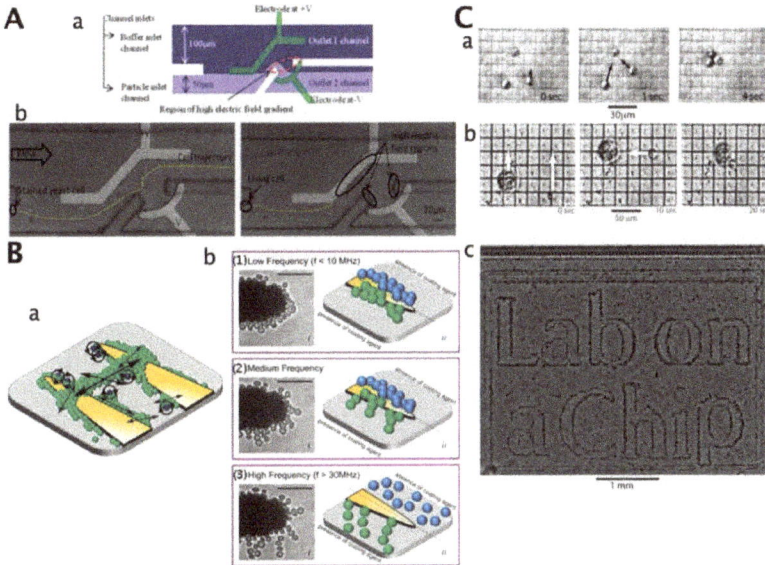

Figure 6. (**A**) (**a**) Schematic of a dielectrophoresis (DEP) sorter with buffer and particle flows being properly balanced; (**b**) Trajectory of a dead yeast cell experiencing negative DEP at the constriction (**left** panel); living yeast cells trapped under positive DEP (**right** panel). Reprinted with permission from ref. [282]. (**B**) (**a**) Three-dimensional schematic of the localised motion of patterned cell clusters under the influence of dielectrophoresis. Cells can exhibit rotational and vibrational movements according to their location within the cluster. Cells located at the free ends of pearl chains have the highest occurrence of rotation, which are indicated as blue cells. Alternatively, the cells packed along the long chains bridged between the microelectrodes have the highest occurrence of vibration. The arrows indicate the direction and strength of the vibrational movement of the cells; (**b**) Response of BSA-treated yeast cell clusters patterned onto a finger-shaped microelectrode array when operated with a 5 Vpk−pk AC sinusoid signal. (1-i, 2-i, 3-i) Response of BSA-treated cell clusters at 5, 20, and 40 MHz, respectively. (1-ii, 2-ii, 3-ii) Schematic representation of the response of patterned cell clusters at different frequencies, with BSA-treated cells (green) being compared to untreated cells (blue). The BSA-treated cells exhibit a positive DEP response, as the distance between the two adjacent cells increases at high frequencies. Alternatively, the untreated cells exhibit a negative DEP response at frequencies higher than 30 MHz. Reprinted with permission from ref. [286]. Copyright (2015) American Chemical Society. (**C**) (**a**) Time sequence of the DEP manipulation of yeast cells. Pixels are energised in sequence to move first one cell alone and then all three together; (**b**) Time sequence of yeast and rat alveolar macrophages manipulated with DEP. Pixels on the chip were energised to independently move the two cells and then bring them together; (**c**) Complex pattern of thousands of yeast cells patterned by DEP. Pixels across the array were energised to spell out "Lab on a Chip", attracting cells toward the local maxima of the electric field. Reprinted with permission from ref. [287].

Table 8. Examples of yeast cell electrical and magnetic manipulation.

Yeast Type	Manipulation Method	Issue Addressed	Refs
S. cerevisiae	Electrophoresis and electroosmosis	Cell transport in microfluidic channels	[272]
	Dielectrophoresis	Live and dead cell separation	[278]
	Magnetic patterning	Demonstration of magnetic micromanipulation of magnetically labelled cells	[276]
	Electroosmosis	Cell transport via electromigration in polymer-based microfluidic devices	[288]
	Dielectrophoresis (AC DEP)	Sorting live and dead cells	[280]
	Dielectrophoresis (AC DEP)	DEP tweezer for single cell manipulation	[267]
	Dielectrophoresis	Multiple frequency DEP separation and trapping of live and dead cells	[281]
	Diamagnetic trapping	Cell magnetic trapping in an array using a CoPt micromagnet array	[289]
	Magnetophoresis	Contactless diamagnetic trapping of cells onto a micromaget array	[289]
	Electrophoresis	Electrophoretic cell manipulation in a microfluidic device	[290]
	Dielectrophoresis (AC DEP)	Separation of yeast cells from blood cells in a microfluidic chip	[268]
	Dielectrophoresis	Live and dead cell separation	[279]
	Dielectrophoresis	Microfluidic chip for guiding cells by AC electrothermal effect and capturing by nDEP trap	[277]
	Dielectrophoresis (DC DEP)	Separation of a mixture of *S. cerevisiae* and *Escherichia coli* cells	[291]
	Dielectrophoresis (AC/DC DEP)	Sorting live and dead cells	[282]
	Dielectrophoresis, electroosmosis, electrophoresis	High-throughput trapping of cells, separation of live and dead cells	[271]
	Dielectrophoresis	Cell manipulation and immobilisation using photo-crosslinkable resin inside microfluidic devices	[292]
	Dielectrophoresis	Controlled rotation and vibration of cell clusters	[286]
	Magnetic manipulation	Magnetic manipulation of Fe_3O_4-doped hydrogel-coated cells	[293]

Magnetophoresis was applied to pattern yeast cells using a micromagnetic array [289]. Therefore, the diamagnetic *S. cerevisiae* were placed in an aqueous solution enriched in paramagnetic ions, and micromagnets that produce high magnetic field gradients were used. *S. cerevisiae* were coated with a single-layer of Fe_3O_4 nanoparticle-doped alginate hydrogel, which allowed their manipulation by a magnetic field [293]. Magnetic and electric manipulation of single or multiple yeast cells in a microfluidic channel was demonstrated using a microelectromagnet matrix and a micropost matrix [276]. The yeast cells labelled with magnetic beads were trapped by the microelectromagnet matrix, whereas the unlabelled cells were trapped by micropost matrix-generating electrical fields. The setup is suitable for the efficient sorting of yeast cells in a microfluidic chip. Yeast cells were trapped in a three-dimensional magnetic trap in an aqueous solution of paramagnetic ions [294].

Magnetic tweezers are similar in concept to optical tweezers; a magnetic particle in an external magnetic field experiences a force proportional to the gradient of the square of the magnetic field [295]. High forces can be achieved with relatively small magnetic field strengths, provided a very steep field gradient can be generated. The fields generated by sharp electromagnetic tips [296] or small permanent magnets [297] have been used to apply forces in excess of 200 pN on micron-sized magnetic particles. Magnetic tweezers are capable of exerting forces in excess of one nN (electromagnetic tweezers), and can be used to manipulate—and importantly, rotate—magnetic particles ranging in size from 0.5 to 5 μm. Magnetic tweezers are unique in that they afford passive, infinite bandwidth, force clamping over large displacements.

4. Conclusions

In recent years, single-molecule and single-cell analysis and manipulation techniques have been developed and applied to the study of yeast cells. Single-cell analysis has increasingly been recognised as the key technology for the elucidation of cellular functions, which are not accessible from bulk measurements at the population level. Various techniques are now available for the analysis of a single cell; with the aid of these techniques, many biological questions can be answered. A microfluidic device is now a suitable technique for single-cell analysis, because a microfluidic system can be manipulated with high throughput, and the amount of sample from a single cell is limited. As it became obvious from this review, the newly developed nanotechniques have been largely applied to the model yeasts *S. cerevisiae* and *Sc. pombe* for fundamental eukaryotic cell biology research, and the pathogenic model yeast *C. albicans* for elucidating the molecular basis of pathogen–host interactions.

High-resolution imaging techniques can provide up to single-biomolecule resolution. The most widely used imaging methods are scanning probe microscopy (i.e., AFM), super-resolution fluorescence microscopy, and electron microscopy. Their characteristics, advantages, and limitations are compared in Table 9. As can be noticed, nanoscale imaging methods are complementary, and they are therefore combined in recently developed imaging platforms, such as bio-AFM and super-resolution fluorescence microscopy [298,299], or the integration of EM and super-resolution microscopy in correlative light and electron microscopy (CLEM) [300–302]. The nanoscale exploration of surfaces of microbes such as yeast cells using AFM has expanded rapidly in the past years. Using AFM topographic imaging, the surface structure of live cells under physiological conditions is achieved with unprecedented resolution. Real-time imaging allows dynamic events to be followed. Chemical force microscopy (CFM)—in which AFM tips are functionalized with specific functional groups—can be used to measure interaction forces on the surface of live yeast cells. Molecular recognition imaging using spatially resolved force spectroscopy, dynamic recognition imaging, or immunogold detection can be used to localize specific receptors, such as yeast adhesins. Quantitative analysis of cell–cell or cell–substrate interactions can be performed with a number of techniques, where AFM single-cell force microscopy, optical tweezers, magnetic tweezers, and micropipette manipulation are the most popular. Understanding the fundamental forces involved in the adhesion of yeast cells is important not only in microbiology, to elucidate cellular functions (such as ligand-binding or biofilm formation), but also in medicine (host-pathogen interactions) and biotechnology (cell aggregation). These force spectroscopy techniques are compared in Table 10. These techniques are complementary, since each technique is most suitable for a specific force range.

Table 9. Comparison of high-resolution techniques for imaging yeast cells (adapted from [303]).

Characteristic	AFM	Electron Microscopy (SEM, TEM)	Super-Resolution Fluorescence Microscopy (PALM, STORM, SIM)
Resolution	~10 nm [1]	~1–10 nm	~5–50 nm
Live cell	Yes	No	Yes
Sample preparation requirement	Little	Little to substantial	Little to moderate
Sample preparation time	10 min–1 d	2 h–5 d	30 min
Image acquisition time	~5 min	5–10 min	Up to 24 h
Equipment cost	€150,000–350,000	€500,000	€250,000–500,000
Operational costs	Low	High	Moderate
Advantages	Localisation (and force spectroscopy) of single proteins; observation of dynamic processes; various environments (temperature, liquid, air, etc.)	Imaging of the cell ultrastructure at very high resolution	Time resolution.
Disadvantages	Only the cell surface is analysed; only one single cell at a time; slow temporal resolution; various sources of artifacts, such as cell or tip alteration	Fixation artifacts; no dynamics; no information on physical properties of proteins	Labelling is required

[1] Depends on the flatness of the surface; the provided value refers to the resolution for observing cells.

Table 10. Comparison of force spectroscopy techniques [295,304].

Characteristic	Optical Tweezers	Magnetic Tweezers	AFM	Micropipette
Type	Point Non-contact	Global/point Non-contact	Point Contact	Point Contact
Spatial resolution (nm)	0.1–2	5–10	0.5–1	-
Temporal resolution (s)	10^{-4}	10^{-1}–10^{-2}	10^{-3}	-
Stiffness (pN nm^{-1})	0.005–1	10^{-3}–10^{-6}	10–10^5	0.01–1000
Force range (pN)	0.1–100	10^{-3}–10^2	10–10^4	1–1000
Probe size (μm)	0.25–5	0.5–5	100–250	
Energy dissipation	Yes	No	No	No
Surface considerations	No	No	Yes	Yes
Features	Low noise and drift dumbbell geometry; access inside a cell	Force clamp, bead rotation, specific interactions; access inside a cell	High-resolution imaging	Controlled deposition/transfer of selected cells
Limitations	Photodamage, sample heating, non specific	No manipulation (force hysteresis)	Large high-stiffness probe, large minimal force, non specific	Low throughput

Several micro-nanomanipulation tools for cells have been developed. The methods can be based on direct-contact mechanical cell manipulation (such as AFM-based or micropipette-based manipulation), or based on non-contact cell manipulation (such as optical, electrical, and magnetic cell manipulation). Examples of these tools are the microchannel-embedded AFM microcantilevers that can be used to suck up one selected yeast cell, which can be further manipulated (positioned for patterning, pushed to another cell or substrate to perform SCFS), and robotic cell printing with picolitre volume dispensing. The manipulation of the physical location of cells is useful both to organise the cells in vitro for single-cell analysis and for specific cell–cell interaction analyses. Another recently developed tool is the use of the AFM cantilever as a very sensitive nanosensor that can detect the metabolic activity of living yeast cells, and even monitor protein conformational changes [305]. An optical tweezer can be used to manipulate several cells in 3D in a contactless way, and can also be applied as a micro-nanosurgery tool by using the nano-ablation option of the laser. Inside cell manipulation of structures has been demonstrated for optical and magnetic tweezers, and opens new possibilities for non-invasive cell organelle manipulation activities. Magnetic tweezers allow cell rotation, which can be important for cell surface location-dependent interactions (e.g., cell–cell interaction analysis during mating).

Electric and magnetic force can be used to trap and position cells at some physical location, to monitor cell viability and separate live from dead cells, transport cells in devices such as lab-on-a-chip to develop automated assays, and to characterise cell properties (e.g., by determining dielectrophoretic spectra of cells).

Acknowledgments: The Belgian Federal Science Policy Office (Belspo) and the European Space Agency (ESA) PRODEX program supported this work. The Research Council of the Vrije Universiteit Brussel (Belgium) and the University of Ghent (Belgium) are acknowledged to support the Alliance Research Group VUB-UGent NanoMicrobiology (NAMI), and the International Joint Research Group (IJRG) VUB-EPFL BioNanotechnology & NanoMedicine (NANO).

Conflicts of Interest: The authors declare no conflict of interest.

References

1. Roco, M.C.; Williams, R.S.; Alivisatos, P. (Eds.) *Biological, Medical and Health Applications: Nanotechnology Research Directions*; Kluwer Academic Publishers: Dordrecht, The Netherlands, 2000.
2. Roco, M.C. Nanotechnology: Convergence with modern biology and medicine. *Curr. Opin. Biotechnol.* **2003**, *14*, 337–346. [CrossRef]

3. Whitesides, G.M. The 'right' size in nanobiotechnology. *Nat. Biotechnol.* **2003**, *21*, 1161–1165. [CrossRef] [PubMed]

4. Binnig, G.; Rohrer, H. Scanning tunnelling microscopy. *Helv. Phys. Acta* **1982**, *55*, 726–735.

5. Binnig, G.; Quate, C.F.; Gerber, C. Atomic force microscope. *Phys. Rev. Lett.* **1986**, *56*, 930–933. [CrossRef] [PubMed]

6. Kada, G.; Kienberger, F.; Hinterdorfer, P. Atomic force microscopy in bionanotechnology. *Nano Today* **2008**, *3*, 12–19. [CrossRef]

7. Ando, T. High-speed atomic force microscopy. *Microscopy (Oxf.)* **2013**, *62*, 81–93. [CrossRef] [PubMed]

8. Kasas, S.; Ikai, A. A method for anchoring round shaped cells for atomic force microscope imaging. *Biophys. J.* **1995**, *68*, 1678–1680. [CrossRef]

9. Gad, M.; Ikai, A. Method for immobilizing microbial cells on gel surface for dynamic AFM studies. *Biophys. J.* **1995**, *69*, 2226–2233. [CrossRef]

10. De, T.; Chettoor, A.M.; Agarwal, P.; Salapaka, M.V.; Nettikadan, S. Immobilization method of yeast cells for intermittent contact mode imaging using the atomic force microscope. *Ultramicroscopy* **2010**, *110*, 254–258. [CrossRef] [PubMed]

11. Dague, E.; Jauvert, E.; Laplatine, L.; Viallet, B.; Thibault, C.; Ressier, L. Assembly of live micro-organisms on microstructured PDMS stamps by convective/capillary deposition for AFM bio-experiments. *Nanotechnology* **2011**, *22*, 395102. [CrossRef] [PubMed]

12. Formosa, C.; Pillet, F.; Schiavone, M.; Duval, R.E.; Ressier, L.; Dague, E. Generation of living cell arrays for atomic force microscopy studies. *Nat. Protoc.* **2015**, *10*, 199–204. [CrossRef] [PubMed]

13. Kasas, S.; Dietler, G. Probing nanomechanical properties from biomolecules to living cells. *Pflugers Arch.* **2008**, *456*, 13–27. [CrossRef] [PubMed]

14. West, M.; Zurek, N.; Hoenger, A.; Voeltz, G.K. A 3D analysis of yeast ER structure reveals how ER domains are organized by membrane curvature. *J. Cell Biol.* **2011**, *193*, 333–346. [CrossRef] [PubMed]

15. Ries, J.; Kaplan, C.; Platonova, E.; Eghlidi, H.; Ewers, H. A simple, versatile method for GFP-based super-resolution microscopy via nanobodies. *Nat. Methods* **2012**, *9*, 582–584. [CrossRef] [PubMed]

16. Formosa, C.; Schiavone, M.; Martin-Yken, H.; François, J.M.; Duval, R.E.; Dague, E. Nanoscale effects of caspofungin against two yeast species, *Saccharomyces cerevisiae* and *Candida albicans. Antimicrob. Agents Chemother.* **2013**, *57*, 3498–3506. [CrossRef] [PubMed]

17. Chopinet, L.; Formosa, C.; Rols, M.P.; Duval, R.E.; Dague, E. Imaging living cells surface and quantifying its properties at high resolution using AFM in QI™ mode. *Micron* **2013**, *48*, 26–33. [CrossRef] [PubMed]

18. Formosa, C.; Schiavone, M.; Boisrame, A.; Richard, M.L.; Duval, R.E.; Dague, E. Multiparametric imaging of adhesive nanodomains at the surface of *Candida albicans* by atomic force microscopy. *Nanomedicine* **2015**, *11*, 57–65. [CrossRef] [PubMed]

19. Mendez-Vilas, A.; Gallardo, A.M.; Perez-Giraldo, C.; Gonzalez-Martín, M.L.; Nuevo, M.J. Surface morphological characterization of yeast cells by scanning force microscopy. *Surf. Interface Anal.* **2001**, *31*, 1027–1030. [CrossRef]

20. Gad, M.; Itoh, A.; Ikai, A. Mapping cell wall polysaccharides of living microbial cells using atomic force microscopy. *Cell Biol. Int.* **1997**, *21*, 697–706. [CrossRef] [PubMed]

21. Touhami, A.; Nysten, B.; Dufrêne, Y.F. Nanoscale mapping of the elasticity of microbial cells by Atomic Force Microscopy. *Langmuir* **2003**, *19*, 4539–4543. [CrossRef]

22. Adya, A.K.; Canetta, E.; Walker, G.M. Atomic force microscopic study of the influence of physical stresses on *Saccharomyces cerevisiae* and *Schizosaccharomyces pombe. FEMS Yeast Res.* **2006**, *6*, 120–128. [CrossRef] [PubMed]

23. Pelling, A.E.; Sehati, S.; Gralla, E.B.; Gimzewski, J.K. Time dependence of the frequency and amplitude of the local nanomechanical motion of yeast. *Nanomedicine* **2005**, *1*, 178–183. [CrossRef] [PubMed]

24. Voychuk, S.I.; Gromozova, E.N.; Lytvyn, P.M.; Podgorsky, V.S. Changes of surface properties of yeast cell wall under exposure of electromagnetic field (40.68 MHz) and action of nystatin. *Environmentalist* **2005**, *25*, 139–144. [CrossRef]

25. Stephens, D.J.; Allan, V.J. Light microscopy techniques for live cell imaging. *Science* **2003**, *300*, 82–86. [CrossRef] [PubMed]

26. Roy, P.; Rajfur, Z.; Pomorski, P.; Jacobson, K. Microscope-based techniques to study cell adhesion and migration. *Nat. Cell Biol.* **2002**, *4*, E91–E96. [CrossRef] [PubMed]

27. Gitai, Z. New fluorescence microscopy methods for microbiology: Sharper, faster, and quantitative. *Curr. Opin. Microbiol.* **2009**, *12*, 341–346. [CrossRef] [PubMed]

28. Fricker, M.; Runions, J.; Moore, I. Quantitative fluorescence microscopy: From art to science. *Annu. Rev. Plant Biol.* **2006**, *57*, 79–107. [CrossRef] [PubMed]

29. Lippincott-Schwartz, J.; Snapp, E.; Kenworthy, A. Studying protein dynamics in living cells. *Nat. Rev. Mol. Cell Biol.* **2001**, *2*, 444–456. [CrossRef] [PubMed]

30. Jares-Erijman, E.A.; Jovin, T.M. FRET imaging. *Nat. Biotechnol.* **2003**, *21*, 1387–1395. [CrossRef] [PubMed]

31. Roy, R.; Hohng, S.; Ha, T. A practical guide to single-molecule FRET. *Nat. Methods* **2008**, *5*, 507–516. [CrossRef] [PubMed]

32. Strutt, J.W. On the manufacture and theory of diffraction-gratings. *Philos. Mag.* **1874**, *47*, 193–205.

33. Thompson, R.E.; Larson, D.R.; Webb, W.W. Precise nanometer localization analysis for individual fluorescent probes. *Biophys. J.* **2002**, *82*, 2775–2783. [CrossRef]

34. Sako, Y. Imaging single molecules in living cells for systems biology. *Mol. Syst. Biol.* **2006**, *2*, 56. [CrossRef] [PubMed]

35. Walter, N.G.; Huang, C.Y.; Manzo, A.J.; Sobhy, M.A. Do-it-yourself guide: How to use the modern single-molecule toolkit. *Nat. Methods* **2008**, *5*, 475–489. [CrossRef] [PubMed]

36. Dehmelt, L.; Bastiaens, P.I. Spatial organization of intracellular communication: Insights from imaging. *Nat. Rev. Mol. Cell Biol.* **2010**, *11*, 440–452. [CrossRef] [PubMed]

37. Harriss, L.M.; Wallace, M.I. Single molecule fluorescence in membrane biology. In *Single Molecule Biology*; Knight, A.E., Ed.; Academic Press: San Diego, CA, USA, 2009; pp. 253–288.

38. Müller-Taubenberger, A.; Anderson, K.I. Recent advances using green and red fluorescent protein variants. *Appl. Microbiol. Biotechnol.* **2007**, *77*, 1–12. [CrossRef] [PubMed]

39. Ando, R.; Hama, H.; Yamamoto-Hino, M.; Mizuno, H.; Miyawaki, A. An optical marker based on the UV-induced green-to-red photoconversion of a fluorescent protein. *Proc. Natl. Acad. Sci. USA* **2002**, *99*, 12651–12656. [CrossRef] [PubMed]

40. Patterson, G.H.; Lippincott-Schwartz, J. A photoactivatable GFP for selective photolabeling of proteins and cells. *Science* **2002**, *297*, 1873–1877. [CrossRef] [PubMed]

41. Chudakov, D.M.; Belousov, V.V.; Zaraisky, A.G.; Novoselov, V.V.; Staroverov, D.B.; Zorov, D.B.; Lukyanov, S.; Lukyanov, K.A. Kindling fluorescent proteins for precise in vivo photolabeling. *Nat. Biotechnol.* **2003**, *21*, 191–194. [CrossRef] [PubMed]

42. Lippincott-Schwartz, J.; Patterson, G.H. Development and use of fluorescent protein markers in living cells. *Science* **2003**, *300*, 87–91. [CrossRef] [PubMed]

43. Chudakov, D.M.; Verkhusha, V.V.; Staroverov, D.B.; Souslova, E.A.; Lukyanov, S.; Lukyanov, K.A. Photoswitchable cyan fluorescent protein for protein tracking. *Nat. Biotechnol.* **2004**, *22*, 1435–1439. [CrossRef] [PubMed]

44. Wiedenmann, J.; Ivanchenko, S.; Oswald, F.; Schmitt, F.; Röcker, C.; Salih, A.; Spindler, K.D.; Nienhaus, G.U. EosFP, a fluorescent marker protein with UV-inducible green-to-red fluorescence conversion. *Proc. Natl. Acad. Sci. USA* **2004**, *101*, 15905–15910. [CrossRef] [PubMed]

45. Gurskaya, N.G.; Verkhusha, V.V.; Shcheglov, A.S.; Staroverov, D.B.; Chepurnykh, T.V.; Fradkov, A.F.; Lukyanov, S.; Lukyanov, K.A. Engineering of a monomeric green-to-red photoactivatable fluorescent protein induced by blue light. *Nat. Biotechnol.* **2006**, *24*, 461–465. [CrossRef] [PubMed]

46. Wiedenmann, J.; Nienhaus, G.U. Live-cell imaging with EosFP and other photoactivatable marker proteins of the GFP family. *Expert Rev. Proteom.* **2006**, *3*, 361–374. [CrossRef] [PubMed]

47. Lippincott-Schwartz, J.; Patterson, G.H. Fluorescent proteins for photoactivation experiments. *Methods Cell Biol.* **2008**, *85*, 45–61. [PubMed]

48. Patterson, G.H.; Lippincott-Schwartz, J. Selective photolabeling of proteins using photoactivatable GFP. *Methods* **2004**, *32*, 445–450. [CrossRef] [PubMed]

49. Habuchi, S.; Ando, R.; Dedecker, P.; Verheijen, W.; Mizuno, H.; Miyawaki, A.; Hofkens, J. Reversible single-molecule photoswitching in the GFP-like fluorescent protein Dronpa. *Proc. Natl. Acad. Sci. USA* **2005**, *102*, 9511–9516. [CrossRef] [PubMed]

50. Verkhusha, V.V.; Sorkin, A. Conversion of the monomeric red fluorescent protein into a photoactivatable probe. *Chem. Biol.* **2005**, *12*, 279–285. [CrossRef] [PubMed]

51. Vogt, A.; D'Angelo, C.; Oswald, F.; Denzel, A.; Mazel, C.H.; Matz, M.V.; Ivanchenko, S.; Nienhaus, G.U.; Wiedenmann, J. A green fluorescent protein with photoswitchable emission from the deep sea. *PLoS ONE* **2008**, *3*, e3766. [CrossRef] [PubMed]
52. Bourgeois, D.; Adam, V. Reversible photoswitching in fluorescent proteins: A mechanistic view. *IUBMB Life* **2012**, *64*, 482–491. [CrossRef] [PubMed]
53. Zhou, X.X.; Lin, M.Z. Photoswitchable fluorescent proteins: Ten years of colorful chemistry and exciting applications. *Curr. Opin. Chem. Biol.* **2013**, *17*, 682–690. [CrossRef] [PubMed]
54. Duan, C.; Adam, V.; Byrdin, M.; Bourgeois, D. Structural basis of photoswitching in fluorescent proteins. *Methods Mol. Biol.* **2014**, *1148*, 177–202. [PubMed]
55. Nienhaus, G.U.; Nienhaus, K.; Hölzle, A.; Ivanchenko, S.; Renzi, F.; Oswald, F.; Wolff, M.; Schmitt, F.; Röcker, C.; Vallone, B.; et al. Photoconvertible fluorescent protein EosFP: Biophysical properties and cell biology applications. *Photochem. Photobiol.* **2006**, *82*, 351–358. [CrossRef] [PubMed]
56. Adam, V.; Nienhaus, K.; Bourgeois, D.; Nienhaus, G.U. Structural basis of enhanced photoconversion yield in green fluorescent protein-like protein Dendra2. *Biochemistry* **2009**, *48*, 4905–4915. [CrossRef] [PubMed]
57. Lukyanov, K.A.; Chudakov, D.M.; Lukyanov, S.; Verkhusha, V.V. Innovation: Photoactivatable fluorescent proteins. *Nat. Rev. Mol. Cell Biol.* **2005**, *6*, 885–891. [CrossRef] [PubMed]
58. Zhang, L.; Gurskaya, N.G.; Merzlyak, E.M.; Staroverov, D.B.; Mudrik, N.N.; Samarkina, O.N.; Vinokurov, L.M.; Lukyanov, S.; Lukyanov, K.A. Method for real-time monitoring of protein degradation at the single cell level. *Biotechniques* **2007**, *42*, 446, 448, 450. [CrossRef] [PubMed]
59. Patterson, G.H. Photoactivation and imaging of optical highlighter fluorescent proteins. *Curr. Protoc. Cytom.* **2011**. [CrossRef]
60. Manley, S.; Gillette, J.M.; Patterson, G.H.; Shroff, H.; Hess, H.F.; Betzig, E.; Lippincott-Schwartz, J. High-density mapping of single-molecule trajectories with photoactivated localization microscopy. *Nat. Methods* **2008**, *5*, 155–157. [CrossRef] [PubMed]
61. Gould, T.J.; Verkhusha, V.V.; Hess, S.T. Imaging biological structures with fluorescence photoactivation localization microscopy. *Nat. Protoc.* **2009**, *4*, 291–308. [CrossRef] [PubMed]
62. Greenfield, D.; McEvoy, A.L.; Shroff, H.; Crooks, G.E.; Wingreen, N.S.; Betzig, E.; Liphardt, J. Self-organization of the *Escherichia coli* chemotaxis network imaged with super-resolution light microscopy. *PLoS Biol.* **2009**, *7*, e1000137. [CrossRef] [PubMed]
63. Betzig, E.; Patterson, G.H.; Sougrat, R.; Lindwasser, O.W.; Olenych, S.; Bonifacino, J.S.; Davidson, M.W.; Lippincott-Schwartz, J.; Hess, H.F. Imaging intracellular fluorescent proteins at nanometer resolution. *Science* **2006**, *313*, 1642–1645. [CrossRef] [PubMed]
64. Rust, M.J.; Bates, M.; Zhuang, X. Sub-diffraction-limit imaging by stochastic optical reconstruction microscopy (STORM). *Nat. Methods* **2006**, *3*, 793–795. [CrossRef] [PubMed]
65. Flors, C.; Hotta, J.; Uji-I, H.; Dedecker, P.; Ando, R.; Mizuno, H.; Miyawaki, A.; Hofkens, J. A stroboscopic approach for fast photoactivation-localization microscopy with Dronpa mutants. *J. Am. Chem. Soc.* **2007**, *129*, 13970–13977. [CrossRef] [PubMed]
66. Hell, S.W. Far-field optical nanoscopy. *Science* **2007**, *316*, 1153–1158. [CrossRef] [PubMed]
67. Stiel, A.C.; Andresen, M.; Bock, H.; Hilbert, M.; Schilde, J.; Schönle, A.; Eggeling, C.; Egner, A.; Hell, S.W.; Jakobs, S. Generation of monomeric reversibly switchable red fluorescent proteins for far-field fluorescence nanoscopy. *Biophys. J.* **2008**, *95*, 2989–2997. [CrossRef] [PubMed]
68. Deschout, H.; Shivanandan, A.; Annibale, P.; Scarselli, M.; Radenovic, A. Progress in quantitative single-molecule localization microscopy. *Histochem. Cell Biol.* **2014**, *142*, 5–17. [CrossRef] [PubMed]
69. Huang, B.; Wang, W.; Bates, M.; Zhuang, X. Three-dimensional super-resolution imaging by stochastic optical reconstruction microscopy. *Science* **2008**, *319*, 810–813. [CrossRef] [PubMed]
70. Juette, M.F.; Gould, T.J.; Lessard, M.D.; Mlodzianoski, M.J.; Nagpure, B.S.; Bennett, B.T.; Hess, S.T.; Bewersdorf, J. Three-dimensional sub-100 nm resolution fluorescence microscopy of thick samples. *Nat. Methods* **2008**, *5*, 527–529. [CrossRef] [PubMed]
71. Jensen, E.; Crossman, D.J. Technical review: Types of imaging-direct STORM. *Anat. Rec. (Hoboken)* **2014**, *297*, 2227–2231. [CrossRef] [PubMed]
72. Tam, J.; Merino, D. Stochastic optical reconstruction microscopy (STORM) in comparison with stimulated emission depletion (STED) and other imaging methods. *J. Neurochem.* **2015**, *135*, 643–658. [CrossRef] [PubMed]

73. Bock, H.; Geisler, C.; Wurm, C.A.; von Middendorf, C.; Jakobs, S.; Schönle, A.; Egner, A.; Hell, S.W.; Eggeling, C. Two-color far-field fluorescence nanoscopy based on photoswitchable emitters. *Appl. Phys. B* **2007**, *88*, 161–165. [CrossRef]

74. Egner, A.; Geisler, C.; von Middendorff, C.; Bock, H.; Wenzel, D.; Medda, R.; Andresen, M.; Stiel, A.C.; Jakobs, S.; Eggeling, C.; et al. Fluorescence nanoscopy in whole cells by asynchronous localization of photoswitching emitters. *Biophys. J.* **2007**, *93*, 3285–3290. [CrossRef] [PubMed]

75. Hess, S.T.; Girirajan, T.P.; Mason, M.D. Ultra-high resolution imaging by fluorescence photoactivation localization microscopy. *Biophys. J.* **2006**, *91*, 4258–4272. [CrossRef] [PubMed]

76. Shroff, H.; Galbraith, C.G.; Galbraith, J.A.; White, H.; Gillette, J.; Olenych, S.; Davidson, M.W.; Betzig, E. Dual-color superresolution imaging of genetically expressed probes within individual adhesion complexes. *Proc. Natl. Acad. Sci. USA* **2007**, *104*, 20308–20313. [CrossRef] [PubMed]

77. Klar, T.A.; Jakobs, S.; Dyba, M.; Egner, A.; Hell, S.W. Fluorescence microscopy with diffraction resolution barrier broken by stimulated emission. *Proc. Natl. Acad. Sci. USA* **2000**, *97*, 8206–8210. [CrossRef] [PubMed]

78. Donnert, G.; Keller, J.; Medda, R.; Andrei, M.A.; Rizzoli, S.O.; Lührmann, R.; Jahn, R.; Eggeling, C.; Hell, S.W. Macromolecular-scale resolution in biological fluorescence microscopy. *Proc. Natl. Acad. Sci. USA* **2006**, *103*, 11440–11445. [CrossRef] [PubMed]

79. Willig, K.I.; Rizzoli, S.O.; Westphal, V.; Jahn, R.; Hell, S.W. STED microscopy reveals that synaptotagmin remains clustered after synaptic vesicle exocytosis. *Nature* **2006**, *440*, 935–939. [CrossRef] [PubMed]

80. Westphal, V.; Rizzoli, S.O.; Lauterbach, M.A.; Kamin, D.; Jahn, R.; Hell, S.W. Video-rate far-field optical nanoscopy dissects synaptic vesicle movement. *Science* **2008**, *320*, 246–249. [CrossRef] [PubMed]

81. De Jonge, N.; Peckys, D.B.; Kremers, G.J.; Piston, D.W. Electron microscopy of whole cells in liquid with nanometer resolution. *Proc. Natl. Acad. Sci. USA* **2009**, *106*, 2159–2164. [CrossRef] [PubMed]

82. Peckys, D.B.; de Jonge, N. Liquid scanning transmission electron microscopy: Imaging protein complexes in their native environment in whole eukaryotic cells. *Microsc. Microanal.* **2014**, *20*, 346–365. [CrossRef] [PubMed]

83. Baumeister, W.; Grimm, R.; Walz, J. Electron tomography of molecules and cells. *Trends Cell Biol.* **1999**, *9*, 81–85. [CrossRef]

84. Downing, K.H.; Sui, H.; Auer, M. Electron tomography: A 3D view of the subcellular world. *Anal. Chem.* **2007**, *79*, 7949–7957. [CrossRef] [PubMed]

85. Stahlberg, H.; Walz, T. Molecular electron microscopy: State of the art and current challenges. *ACS Chem. Biol.* **2008**, *3*, 268–281. [CrossRef] [PubMed]

86. Lucić, V.; Leis, A.; Baumeister, W. Cryo-electron tomography of cells: Connecting structure and function. *Histochem. Cell Biol.* **2008**, *130*, 185–196. [CrossRef] [PubMed]

87. Diebolder, C.A.; Koster, A.J.; Koning, R.I. Pushing the resolution limits in cryo electron tomography of biological structures. *J. Microsc.* **2012**, *248*, 1–5. [CrossRef] [PubMed]

88. Agar, H.D.; Douglas, H.C. Studies on the cytological structure of yeast: Electron microscopy of thin sections. *J. Bacteriol.* **1957**, *73*, 365–375. [PubMed]

89. Matile, P.; Moor, H.; Robinow, C.F. Yeast cytology. In *The Yeasts*; Rose, A.H., Harrison, J.S., Eds.; Avademic Press: New York, NY, USA, 1969; pp. 219–302.

90. Osumi, M. Visualization of yeast cells by electron microscopy. *J. Electron. Microsc. (Tokyo)* **2012**, *61*, 343–365. [CrossRef] [PubMed]

91. Osumi, M.; Sando, N. Division of yeast mitochondria in synchronous culture. *J. Electron. Microsc. (Tokyo)* **1969**, *18*, 47–56. [PubMed]

92. Byers, B.; Goetsch, L. Duplication of spindle plaques and integration of the yeast cell cycle. *Cold Spring Harb. Symp. Quant. Biol.* **1974**, *38*, 123–131. [CrossRef] [PubMed]

93. Byers, B.; Goetsch, L. Behavior of spindles and spindle plaques in the cell cycle and conjugation of *Saccharomyces cerevisiae*. *J. Bacteriol.* **1975**, *124*, 511–523. [PubMed]

94. Osumi, M.; Imaizumi, F.; Imai, M.; Sato, H.; Yamaguchi, H. Isolation and characterization of microbodies from *candida tropicalis* pk 233 cells grown on normal alkanes. *J. Gen. Appl. Microbiol.* **1975**, *21*, 375–387. [CrossRef]

95. Novick, P.; Field, C.; Schekman, R. Identification of 23 complementation groups required for post-translational events in the yeast secretory pathway. *Cell* **1980**, *21*, 205–215. [CrossRef]

96. Winey, M.; Goetsch, L.; Baum, P.; Byers, B. MPS1 and MPS2: Novel yeast genes defining distinct steps of spindle pole body duplication. *J. Cell Biol.* **1991**, *114*, 745–754. [CrossRef] [PubMed]

97. Baba, M.; Osumi, M. Transmission and scanning electron microscopic examination of intracellular organelles in freeze-substituted *Kloeckera* and *Saccharomyces cerevisiae* yeast cells. *J. Electron. Microsc. Tech.* **1987**, *5*, 249–261. [CrossRef]

98. Giddings, T.H., Jr.; O'Toole, E.T.; Morphew, M.; Mastronarde, D.N.; McIntosh, J.R.; Winey, M. Using rapid freeze and freeze-substitution for the preparation of yeast cells for electron microscopy and three-dimensional analysis. *Methods Cell Biol.* **2001**, *67*, 27–42. [PubMed]

99. McDonald, K. Cryopreparation methods for electron microscopy of selected model systems. *Methods Cell Biol.* **2007**, *79*, 23–56. [PubMed]

100. O'Toole, E.T.; Giddings, T.H., Jr.; Winey, M. Building cell structures in three dimensions: Electron tomography methods for budding yeast. In *Yeast Protocols*; CSHL Press: Cold Spring Harbor, NY, USA, 2016; pp. 303–312.

101. Hoenger, A.; McIntosh, J.R. Probing the macromolecular organization of cells by electron tomography. *Curr. Opin. Cell Biol.* **2009**, *21*, 89–96. [CrossRef] [PubMed]

102. Arfsten, J.; Leupold, S.; Bradtmöller, C.; Kampen, I.; Kwade, A. Atomic force microscopy studies on the nanomechanical properties of *Saccharomyces cerevisiae*. *Colloids Surf. B Biointerfaces* **2010**, *79*, 284–290. [CrossRef] [PubMed]

103. Alsteens, D.; Beaussart, A.; Derclaye, S.; El-Kirat-Chatel, S.; Park, H.R.; Lipke, P.N.; Dufrêne, Y.F. Single-cell force spectroscopy of Als-mediated fungal adhesion. *Anal. Methods* **2013**, *5*, 3657–3662. [CrossRef] [PubMed]

104. Portillo, A.M.; Krasnoslobodtsev, A.V.; Lyubchenko, Y.L. Effect of electrostatics on aggregation of prion protein Sup35 peptide. *J. Phys. Condens. Matter* **2012**, *24*, 164205. [CrossRef] [PubMed]

105. Portillo, A.; Hashemi, M.; Zhang, Y.; Breydo, L.; Uversky, V.N.; Lyubchenko, Y.L. Role of monomer arrangement in the amyloid self-assembly. *Biochim. Biophys. Acta* **2015**, *1854*, 218–228. [CrossRef] [PubMed]

106. Rangl, M.; Ebner, A.; Yamada, J.; Rankl, C.; Tampé, R.; Gruber, H.J.; Rexach, M.; Hinterdorfer, P. Single-molecule analysis of the recognition forces underlying nucleo-cytoplasmic transport. *Angew. Chem. Int. Ed. Engl.* **2013**, *52*, 10356–10359. [CrossRef] [PubMed]

107. Goossens, K.V.Y.; Ielasi, F.S.; Nookaew, I.; Stals, I.; Alonso-Sarduy, L.; Daenen, L.; van Mulders, S.E.; Stassen, C.; van Eijsden, R.G.E.; Siewers, V.; et al. Molecular mechanism of flocculation self-recognition in yeast and its role in mating and survival. *mBio* **2015**, *6*, e00427–e00415. [CrossRef] [PubMed]

108. El-Kirat-Chatel, S.; Beaussart, A.; Alsteens, D.; Sarazin, A.; Jouault, T.; Dufrêne, Y.F. Single-molecule analysis of the major glycopolymers of pathogenic and non-pathogenic yeast cells. *Nanoscale* **2013**, *5*, 4855–4863. [CrossRef] [PubMed]

109. Hwang, G.; Marsh, G.; Gao, L.; Waugh, R.; Koo, H. Binding force dynamics of *Streptococcus mutans*-glucosyltransferase B to *Candida albicans*. *J. Dent. Res.* **2015**, *94*, 1310–1317. [CrossRef] [PubMed]

110. El-Kirat-Chatel, S.; Beaussart, A.; Derclaye, S.; Alsteens, D.; Kuchaříková, S.; van Dijck, P.; Dufrêne, Y.F. Force nanoscopy of hydrophobic interactions in the fungal pathogen *Candida glabrata*. *ACS Nano* **2015**, *9*, 1648–1655. [CrossRef] [PubMed]

111. Touhami, A.; Hoffmann, B.; Vasella, A.; Denis, F.A.; Dufrêne, Y.F. Aggregation of yeast cells: Direct measurement of discrete lectin-carbohydrate interactions. *Microbiology* **2003**, *149*, 2873–2878. [CrossRef] [PubMed]

112. Dupres, V.; Alsteens, D.; Wilk, S.; Hansen, B.; Heinisch, J.J.; Dufrêne, Y.F. The yeast Wsc1 cell surface sensor behaves like a nanospring in vivo. *Nat. Chem. Biol.* **2009**, *5*, 857–862. [CrossRef] [PubMed]

113. Dupres, V.; Heinisch, J.J.; Dufrêne, Y.F. Atomic force microscopy demonstrates that disulphide bridges are required for clustering of the yeast cell wall integrity sensor Wsc1. *Langmuir* **2011**, *27*, 15129–15134. [CrossRef] [PubMed]

114. Heinisch, J.J.; Dupres, V.; Wilk, S.; Jendretzki, A.; Dufrêne, Y.F. Single-molecule atomic force microscopy reveals clustering of the yeast plasma-membrane sensor Wsc1. *PLoS ONE* **2010**, *5*, e11104. [CrossRef] [PubMed]

115. Formosa, C.; Lachaize, V.; Galés, C.; Rols, M.P.; Martin-Yken, H.; François, J.M.; Duval, R.E.; Dague, E. Mapping HA-tagged protein at the surface of living cells by atomic force microscopy. *J. Mol. Recognit.* **2015**, *28*, 1–9. [CrossRef] [PubMed]

116. Takenaka, M.; Miyachi, Y.; Ishii, J.; Ogino, C.; Kondo, A. The mapping of yeast's G-protein coupled receptor with an atomic force microscope. *Nanoscale* **2015**, *7*, 4956–4963. [CrossRef] [PubMed]

117. Friedrichs, J.; Helenius, J.; Muller, D.J. Quantifying cellular adhesion to extracellular matrix components by single-cell force spectroscopy. *Nat. Protoc.* **2010**, *5*, 1353–1361. [CrossRef] [PubMed]

118. Te Riet, J.; Reinieren-Beeren, I.; Figdor, C.G.; Cambi, A. AFM force spectroscopy reveals how subtle structural differences affect the interaction strength between *Candida albicans* and DC-SIGN. *J. Mol. Recognit.* **2015**, *28*, 687–698. [CrossRef] [PubMed]

119. Benoit, M.; Gabriel, D.; Gerisch, G.; Gaub, H.E. Discrete interactions in cell adhesion measured by single-molecule force spectroscopy. *Nat. Cell Biol.* **2000**, *2*, 313–317. [CrossRef] [PubMed]

120. Helenius, J.; Heisenberg, C.P.; Gaub, H.E.; Muller, D.J. Single-cell force spectroscopy. *J. Cell Sci.* **2008**, *121*, 1785–1791. [CrossRef] [PubMed]

121. Friedrichs, J.; Legate, K.R.; Schubert, R.; Bharadwaj, M.; Werner, C.; Müller, D.J.; Benoit, M. A practical guide to quantify cell adhesion using single-cell force spectroscopy. *Methods* **2013**, *60*, 169–178. [CrossRef] [PubMed]

122. Taubenberger, A.; Cisneros, D.A.; Friedrichs, J.; Puech, P.H.; Muller, D.J.; Franz, C.M. Revealing early steps of alpha2beta1 integrin-mediated adhesion to collagen type I by using single-cell force spectroscopy. *Mol. Biol. Cell* **2007**, *18*, 1634–1644. [CrossRef] [PubMed]

123. Peters, B.M.; Ovchinnikova, E.S.; Krom, B.P.; Schlecht, L.M.; Zhou, H.; Hoyer, L.L.; Busscher, H.J.; van der Mei, H.C.; Jabra-Rizk, M.A.; Shirtliff, M.E. *Staphylococcus aureus* adherence to *Candida albicans* hyphae is mediated by the hyphal adhesin Als3p. *Microbiology* **2012**, *158*, 2975–2986. [CrossRef] [PubMed]

124. Potthoff, E.; Guillaume-Gentil, O.; Ossola, D.; Polesel-Maris, J.; LeibundGut-Landmann, S.; Zambelli, T.; Vorholt, J.A. Rapid and serial quantification of adhesion forces of yeast and Mammalian cells. *PLoS ONE* **2012**, *7*, e52712. [CrossRef] [PubMed]

125. Alsteens, D.; van Dijck, P.; Lipke, P.N.; Dufrêne, Y.F. Quantifying the forces driving cell-cell adhesion in a fungal pathogen. *Langmuir* **2013**, *29*, 13473–13480. [CrossRef] [PubMed]

126. Bowen, W.R.; Lovitt, R.W.; Wright, C.J. Atomic Force Microscopy study of the adhesion of *Saccharomyces cerevisiae*. *J. Colloid Interface Sci.* **2001**, *237*, 54–61. [CrossRef] [PubMed]

127. Götzinger, M.; Weigl, B.; Peukert, W.; Sommer, K. Effect of roughness on particle adhesion in aqueous solutions: A study of *Saccharomyces cerevisiae* and a silica particle. *Colloids Surf. B Biointerfaces* **2007**, *55*, 44–50. [CrossRef] [PubMed]

128. Hachułka, K.; Lekka, M.; Okrajni, J.; Ambroziak, W.; Wandelt, B. Polymeric sensing system molecularly imprinted towards enhanced adhesion of *Saccharomyces cerevisiae*. *Biosens. Bioelectron.* **2010**, *26*, 50–54. [CrossRef] [PubMed]

129. Martinez, V.; Behr, P.; Drechsler, U.; Polesel-Maris, J.; Potthoff, E.; Vörös, J.; Zambelli, T. SU-8 hollow cantilevers for AFM cell adhesion studies. *J. Micromech. Microeng.* **2016**, *26*, 055006. [CrossRef]

130. Hansen, K.M.; Thundat, T. Microcantilever biosensors. *Methods* **2005**, *37*, 57–64. [CrossRef] [PubMed]

131. Fritz, J. Cantilever biosensors. *Analyst* **2008**, *133*, 855–863. [CrossRef] [PubMed]

132. Braun, T.; Ghatkesar, M.K.; Backmann, N.; Grange, W.; Boulanger, P.; Letellier, L.; Lang, H.P.; Bietsch, A.; Gerber, C.; Hegner, M. Quantitative time-resolved measurement of membrane protein-ligand interactions using microcantilever array sensors. *Nat. Nanotechnol.* **2009**, *4*, 179–185. [CrossRef] [PubMed]

133. Ndieyira, J.W.; Watari, M.; Barrera, A.D.; Zhou, D.; Vögtli, M.; Batchelor, M.; Cooper, M.A.; Strunz, T.; Horton, M.A.; Abell, C.; et al. Nanomechanical detection of antibiotic-mucopeptide binding in a model for superbug drug resistance. *Nat. Nanotechnol.* **2008**, *3*, 691–696. [CrossRef] [PubMed]

134. Godin, M.; Delgado, F.F.; Son, S.; Grover, W.H.; Bryan, A.K.; Tzur, A.; Jorgensen, P.; Payer, K.; Grossman, A.D.; Kirschner, M.W.; et al. Using buoyant mass to measure the growth of single cells. *Nat. Methods* **2010**, *7*, 387–390. [CrossRef] [PubMed]

135. Lang, H.P.; Baller, M.K.; Berger, R.; Gerber, C.; Gimzewski, J.K.; Battiston, F.M.; Fornaro, P.; Ramseyer, J.P.; Meyer, E.; Guntherodt, H.J. An artificial nose based on a micromechancial cantilever array. *Anal. Chim. Acta* **1999**, *393*, 59–65. [CrossRef]

136. Braun, T.; Barwich, V.; Ghatkesar, M.K.; Bredekamp, A.H.; Gerber, C.; Hegner, M.; Lang, H.P. Micromechanical mass sensors for biomolecular detection in a physiological environment. *Phys. Rev. E Stat. Nonlinear Soft Matter Phys.* **2005**, *72*, 031907. [CrossRef] [PubMed]

137. Hosaka, S.; Chiyoma, T.; Ikeuchi, A.; Okano, H.; Sone, H.; Izumi, T. Possibility of a femtogram mass biosensor using a self-sensing cantilever. *Curr. Appl. Phys.* **2006**, *6*, 384–388. [CrossRef]

138. Liu, Y.; Schweizerb, L.M.; Wanga, W.; Reubena, R.L.; Schweizer, M.; Shu, W. Label-free and real-time monitoring of yeast cell growth by the bending of polymer microcantilever biosensors. *Sens. Actuator B Chem.* **2013**, *178*, 621–626. [CrossRef]

139. Bryan, A.K.; Goranov, A.; Amon, A.; Manalis, S.R. Measurement of mass, density, and volume during the cell cycle of yeast. *Proc. Natl. Acad. Sci. USA* **2010**, *107*, 999–1004. [CrossRef] [PubMed]
140. Godin, M.; Tabard-Cossa, V.; Miyahara, Y.; Monga, T.; Williams, P.J.; Beaulieu, L.Y.; Bruce Lennox, R.; Grutter, P. Cantilever-based sensing: The origin of surface stress and optimization strategies. *Nanotechnology* **2010**, *21*, 75501. [CrossRef] [PubMed]
141. Burg, T.P.; Godin, M.; Knudsen, S.M.; Shen, W.; Carlson, G.; Foster, J.S.; Babcock, K.; Manalis, S.R. Weighing of biomolecules, single cells and single nanoparticles in fluid. *Nature* **2007**, *446*, 1066–1069. [CrossRef] [PubMed]
142. Park, K.; Jang, J.; Irimia, D.; Sturgis, J.; Lee, J.; Robinson, J.P.; Toner, M.; Bashir, R. 'Living cantilever arrays' for characterization of mass of single live cells in fluids. *Lab Chip* **2008**, *8*, 1034–1041. [CrossRef] [PubMed]
143. Bryan, A.K.; Hecht, V.C.; Shen, W.; Payer, K.; Grover, W.H.; Manalis, S.R. Measuring single cell mass, volume, and density with dual suspended microchannel resonators. *Lab Chip* **2014**, *14*, 569–576. [CrossRef] [PubMed]
144. Nugaeva, N.; Gfeller, K.Y.; Backmann, N.; Lang, H.P.; Düggelin, M.; Hegner, M. Micromechanical cantilever array sensors for selective fungal immobilization and fast growth detection. *Biosens. Bioelectron.* **2005**, *21*, 849–856. [CrossRef] [PubMed]
145. Aghayee, S.; Benadiba, C.; Notz, J.; Kasas, S.; Dietler, G.; Longo, G. Combination of fluorescence microscopy and nanomotion detection to characterize bacteria. *J. Mol. Recognit.* **2013**, *26*, 590–595. [CrossRef] [PubMed]
146. Longo, G.; Alonso-Sarduy, L.; Rio, L.M.; Bizzini, A.; Trampuz, A.; Notz, J.; Dietler, G.; Kasas, S. Rapid detection of bacterial resistance to antibiotics using AFM cantilevers as nanomechanical sensors. *Nat. Nanotechnol.* **2013**, *8*, 522–526. [CrossRef] [PubMed]
147. Kasas, S.; Ruggeri, F.S.; Benadiba, C.; Maillard, C.; Stupar, P.; Tournu, H.; Dietler, G.; Longo, G. Detecting nanoscale vibrations as signature of life. *Proc. Natl. Acad. Sci. USA* **2015**, *112*, 378–381. [CrossRef] [PubMed]
148. Voldman, J. Engineered systems for the physical manipulation of single cells. *Curr. Opin. Biotechnol.* **2006**, *17*, 532–537. [CrossRef] [PubMed]
149. Voldman, J. Electrical forces for microscale cell manipulation. *Annu. Rev. Biomed. Eng.* **2006**, *8*, 425–454. [CrossRef] [PubMed]
150. Willaert, R.G.; Goossens, K. Microfluidic bioreactors for cellular microarrays. *Fermentation* **2015**, *1*, 38–78. [CrossRef]
151. Guo, L.J. Nanoimprint lithography: Methods and material requirements. *Adv. Mater.* **2007**, *19*, 495–513. [CrossRef]
152. Yap, F.L.; Zhang, Y. Protein and cell micropatterning and its integration with micro/nanoparticles assembly. *Biosens. Bioelectron.* **2007**, *22*, 775–788. [CrossRef] [PubMed]
153. Anselme, K.; Davidson, P.; Popa, A.M.; Giazzon, M.; Liley, M.; Ploux, L. The interaction of cells and bacteria with surfaces structured at the nanometre scale. *Acta Biomater.* **2010**, *6*, 3824–3846. [CrossRef] [PubMed]
154. Qin, D.; Xia, Y.; Whitesides, G.M. Soft lithography for micro- and nanoscale patterning. *Nat. Protoc.* **2010**, *5*, 491–502. [CrossRef] [PubMed]
155. Ekerdt, B.L.; Segalman, R.A.; Schaffer, D.V. Spatial organization of cell-adhesive ligands for advanced cell culture. *Biotechnol. J.* **2013**, *8*, 1411–1423. [CrossRef] [PubMed]
156. Singh, A.V.; Patil, R.; Thombre, D.K.; Gade, W.N. Micro-nanopatterning as tool to study the role of physicochemical properties on cell-surface interactions. *J. Biomed. Mater. Res. A* **2013**, *101*, 3019–3032. [CrossRef] [PubMed]
157. Fung, T.H.; Ball, G.I.; McQuaide, S.C.; Chao, S.; Colman-Lerner, A.; Holl, M.R.; Meldrum, D.R. Microprinting of on-chip cultures: Patterning of yeast cell microarrays using concanavalin-A adhesion. In Proceedings of the IMECE04 ASME International Mechanical Engineering Congress, Anaheim, CA, USA, 13–19 November 2004; pp. 373–374.
158. Cookson, S.; Ostroff, N.; Pang, W.L.; Volfson, D.; Hasty, J. Monitoring dynamics of single-cell gene expression over multiple cell cycles. *Mol. Syst. Biol.* **2005**, *1*, 0024. [CrossRef] [PubMed]
159. Narayanaswamy, R.; Niu, W.; Scouras, A.D.; Hart, G.T.; Davies, J.; Ellington, A.D.; Iyer, V.R.; Marcotte, E.M. Systematic profiling of cellular phenotypes with spotted cell microarrays reveals mating-pheromone response genes. *Genome Biol.* **2006**, *7*, R6. [CrossRef] [PubMed]
160. Ryley, J.; Pereira-Smith, O.M. Microfluidics device for single cell gene expression analysis in *Saccharomyces cerevisiae*. *Yeast* **2006**, *23*, 1065–1073. [CrossRef] [PubMed]

161. Narayanaswamy, R.; Moradi, E.K.; Niu, W.; Hart, G.T.; Davis, M.; McGary, K.L.; Ellington, A.D.; Marcotte, E.M. Systematic definition of protein constituents along the major polarization axis reveals an adaptive reuse of the polarization machinery in pheromone-treated budding yeast. *J. Proteom. Res.* **2009**, *8*, 6–19. [CrossRef] [PubMed]

162. Falconnet, D.; Niemistö, A.; Taylor, R.J.; Ricicova, M.; Galitski, T.; Shmulevich, I.; Hansen, C.L. High-throughput tracking of single yeast cells in a microfluidic imaging matrix. *Lab Chip* **2011**, *11*, 466–473. [CrossRef] [PubMed]

163. Kuhn, P.; Eyer, K.; Robinson, T.; Schmidt, F.I.; Mercer, J.; Dittrich, P.S. A facile protocol for the immobilisation of vesicles, virus particles, bacteria, and yeast cells. *Integr. Biol. (Camb.)* **2012**, *4*, 1550–1555. [CrossRef] [PubMed]

164. Lee, S.S.; Avalos Vizcarra, I.; Huberts, D.H.; Lee, L.P.; Heinemann, M. Whole lifespan microscopic observation of budding yeast aging through a microfluidic dissection platform. *Proc. Natl. Acad. Sci. USA* **2012**, *109*, 4916–4920. [CrossRef] [PubMed]

165. Park, M.C.; Hur, J.Y.; Cho, H.S.; Park, S.H.; Suh, K.Y. High-throughput single-cell quantification using simple microwell-based cell docking and programmable time-course live-cell imaging. *Lab Chip* **2011**, *11*, 79–86. [CrossRef] [PubMed]

166. Xie, Z.; Zhang, Y.; Zou, K.; Brandman, O.; Luo, C.; Ouyang, Q.; Li, H. Molecular phenotyping of aging in single yeast cells using a novel microfluidic device. *Aging Cell* **2012**, *11*, 599–606. [CrossRef] [PubMed]

167. Zhang, Y.; Luo, C.; Zou, K.; Xie, Z.; Brandman, O.; Ouyang, Q.; Li, H. Single cell analysis of yeast replicative aging using a new generation of microfluidic device. *PLoS ONE* **2012**, *7*, e48275. [CrossRef] [PubMed]

168. Fehrmann, S.; Paoletti, C.; Goulev, Y.; Ungureanu, A.; Aguilaniu, H.; Charvin, G. Aging yeast cells undergo a sharp entry into senescence unrelated to the loss of mitochondrial membrane potential. *Cell Rep.* **2013**, *5*, 1589–1599. [CrossRef] [PubMed]

169. Crane, M.M.; Clark, I.B.; Bakker, E.; Smith, S.; Swain, P.S. A microfluidic system for studying ageing and dynamic single-cell responses in budding yeast. *PLoS ONE* **2014**, *9*, e100042. [CrossRef] [PubMed]

170. Osada, K.; Hosokawa, M.; Yoshino, T.; Tanaka, T. Monitoring of cellular behaviors by microcavity array-based single-cell patterning. *Analyst* **2014**, *139*, 425–430. [CrossRef] [PubMed]

171. Liu, P.; Young, T.Z.; Acar, M. Yeast replicator: A high-throughput multiplexed microfluidics platform for automated measurements of single-cell aging. *Cell Rep.* **2015**, *13*, 634–644. [CrossRef] [PubMed]

172. Jo, M.C.; Liu, W.; Gu, L.; Dang, W.; Qin, L. High-throughput analysis of yeast replicative aging using a microfluidic system. *Proc. Natl. Acad. Sci. USA* **2015**, *112*, 9364–9369. [CrossRef] [PubMed]

173. Terenna, C.R.; Makushok, T.; Velve-Casquillas, G.; Baigl, D.; Chen, Y.; Bornens, M.; Paoletti, A.; Piel, M.; Tran, P.T. Physical mechanisms redirecting cell polarity and cell shape in fission yeast. *Curr. Biol.* **2008**, *18*, 1748–1753. [CrossRef] [PubMed]

174. Minc, N.; Boudaoud, A.; Chang, F. Mechanical forces of fission yeast growth. *Curr. Biol.* **2009**, *19*, 1096–1101. Erratum in: *Curr. Biol.* **2014**, *24*, 1436. [CrossRef] [PubMed]

175. Nghe, P.; Boulineau, S.; Gude, S.; Recouvreux, P.; van Zon, J.S.; Tans, S.J. Microfabricated polyacrylamide devices for the controlled culture of growing cells and developing organisms. *PLoS ONE* **2013**, *8*, e75537. [CrossRef] [PubMed]

176. Tian, Y.; Luo, C.; Ouyang, Q. A microfluidic synchronizer for fission yeast cells. *Lab Chip* **2013**, *13*, 4071–4077. [CrossRef] [PubMed]

177. Bell, L.; Seshia, A.; Lando, D.; Laue, E.; Palayret, M.; Lee, S.F.; Klenerman, D. A microfluidic device for the hydrodynamic immobilisation of living fission yeast cells for super-resolution imaging. *Sens. Actuators B Chem.* **2014**, *192*, 36–41. [CrossRef] [PubMed]

178. Nobs, J.B.; Maerkl, S.J. Long-term single cell analysis of S. *pombe* on a microfluidic microchemostat array. *PLoS ONE* **2014**, *9*, e93466. [CrossRef] [PubMed]

179. Spivey, E.C.; Xhemalce, B.; Shear, J.B.; Finkelstein, I.J. 3D-printed microfluidic microdissector for high-throughput studies of cellular aging. *Anal. Chem.* **2014**, *86*, 7406–7412. [CrossRef] [PubMed]

180. Ruiz, S.A.; Chen, C.S. Microcontact printing: A tool to pattern. *Soft Matter* **2007**, *3*, 168–177. [CrossRef]

181. Miermont, A.; Waharte, F.; Hu, S.; McClean, M.N.; Bottani, S.; Léon, S.; Hersen, P. Severe osmotic compression triggers a slowdown of intracellular signaling, which can be explained by molecular crowding. *Proc. Natl. Acad. Sci. USA* **2013**, *110*, 5725–5730. [CrossRef] [PubMed]

182. Théry, M.; Piel, M. Adhesive micropatterns for cells: A microcontact printing protocol. *Cold Spring Harb. Protoc.* **2009**, *7*. [CrossRef] [PubMed]

183. Weaver, W.M.; Tseng, P.; Kunze, A.; Masaeli, M.; Chung, A.J.; Dudani, J.S.; Kittur, H.; Kulkarni, R.P.; di Carlo, D. Advances in high-throughput single-cell microtechnologies. *Curr. Opin. Biotechnol.* **2014**, *25*, 114–123. [CrossRef] [PubMed]

184. Fritzsch, F.S.; Dusny, C.; Frick, O.; Schmid, A. Single-cell analysis in biotechnology, systems biology, and biocatalysis. *Annu. Rev. Chem. Biomol. Eng.* **2012**, *3*, 129–155. [CrossRef] [PubMed]

185. Dusny, C.; Schmid, A. Microfluidic single-cell analysis links boundary environments and individual microbial phenotypes. *Environ. Microbiol.* **2015**, *17*, 1839–1856. [CrossRef] [PubMed]

186. Groisman, A.; Lobo, C.; Cho, H.; Campbell, J.K.; Dufour, Y.S.; Stevens, A.M.; Levchenko, A. A microfluidic chemostat for experiments with bacterial and yeast cells. *Nat. Methods* **2005**, *2*, 685–689. [CrossRef] [PubMed]

187. Long, Z.; Nugent, E.; Javer, A.; Cicuta, P.; Sclavi, B.; Cosentino Lagomarsino, M.; Dorfman, K.D. Microfluidic chemostat for measuring single cell dynamics in bacteria. *Lab Chip* **2013**, *13*, 947–954. [CrossRef] [PubMed]

188. Balaban, N.Q.; Merrin, J.; Chait, R.; Kowalik, L.; Leibler, S. Bacterial persistence as a phenotypic switch. *Science* **2004**, *305*, 1622–1625. [CrossRef] [PubMed]

189. Bennett, M.R.; Pang, W.L.; Ostroff, N.A.; Baumgartner, B.L.; Nayak, S.; Tsimring, L.S.; Hasty, J. Metabolic gene regulation in a dynamically changing environment. *Nature* **2008**, *454*, 1119–1122. [CrossRef] [PubMed]

190. Yarmush, M.L.; King, K.R. Living-cell microarrays. *Annu. Rev. Biomed. Eng.* **2009**, *11*, 235–257. [CrossRef] [PubMed]

191. Nilsson, J.; Evander, M.; Hammarström, B.; Laurell, T. Review of cell and particle trapping in microfluidic systems. *Anal. Chim. Acta* **2009**, *649*, 141–157. [CrossRef] [PubMed]

192. Di Carlo, D.; Wu, L.Y.; Lee, L.P. Dynamic single cell culture array. *Lab Chip* **2006**, *6*, 1445–1449. [CrossRef] [PubMed]

193. Barbulovic-Nad, I.; Lucente, M.; Sun, Y.; Zhang, M.; Wheeler, A.R.; Bussmann, M. Bio-microarray fabrication techniques—A review. *Crit. Rev. Biotechnol.* **2006**, *26*, 237–259. [CrossRef] [PubMed]

194. Guillemot, F.; Souquet, A.; Catros, S.; Guillotin, B.; Lopez, J.; Faucon, M.; Pippenger, B.; Bareille, R.; Rémy, M.; Bellance, S.; et al. High-throughput laser printing of cells and biomaterials for tissue engineering. *Acta Biomater.* **2010**, *6*, 2494–2500. [CrossRef] [PubMed]

195. Bean, G.J.; Jaeger, P.A.; Bahr, S.; Ideker, T. Development of ultra-high-density screening tools for microbial "omic". *PLoS ONE* **2014**, *9*, e85177. [CrossRef] [PubMed]

196. Schaack, B.; Reboud, J.; Combe, S.; Fouqué, B.; Berger, F.; Boccard, S.; Filhol-Cochet, O.; Chatelain, F. A "DropChip" cell array for DNA and siRNA transfection combined with drug screening. *NanoBiotechnology* **2005**, *1*, 183–189. [CrossRef]

197. Ringeisen, B.R.; Othon, C.M.; Barron, J.A.; Young, D.; Spargo, B.J. Jet-based methods to print living cells. *Biotechnol. J.* **2006**, *1*, 930–948. [CrossRef] [PubMed]

198. Roth, E.A.; Xu, T.; Das, M.; Gregory, C.; Hickman, J.J.; Boland, T. Inkjet printing for high-throughput cell patterning. *Biomaterials* **2004**, *25*, 3707–3715. [CrossRef] [PubMed]

199. Ferris, C.J.; Gilmore, K.G.; Wallace, G.G.; In Het Panhuis, M. Biofabrication: An overview of the approaches used for printing of living cells. *Appl. Microbiol. Biotechnol.* **2013**, *97*, 4243–4258. [CrossRef] [PubMed]

200. Gonzalez-Macia, L.; Morrin, A.; Smyth, M.R.; Killard, A.J. Advanced printing and deposition methodologies for the fabrication of biosensors and biodevices. *Analyst* **2010**, *135*, 845–867. [CrossRef] [PubMed]

201. Li, J.; Rossignol, F.; Macdonald, J. Inkjet printing for biosensor fabrication: Combining chemistry and technology for advanced manufacturing. *Lab Chip* **2015**, *15*, 2538–2558. [CrossRef] [PubMed]

202. Meister, A.; Liley, M.; Brugger, J.; Pugin, R.; Heinzelmann, H. Nanodispenser for attoliter volume deposition using atomic force microscopy probes modified by focused-ion-beam milling. *Appl. Phys. Lett.* **2005**, *85*, 6260–6262. [CrossRef]

203. Deladi, S.; Tas, N.R.; Berenschot, J.W.; de Boer, J.H.; de Boer, M.J.; Peter, M.; Krijnen, G.J.M.; Elwenspoek, M.C. Micromachines fountain pen for atomic force microscope-based nanopatterning. *Appl. Phys. Lett.* **2004**, *85*, 5361. [CrossRef]

204. Kim, K.H.; Moldovan, N.; Espinosa, H.D. A nanofountain probe with sub-100 nm molecular writing resolution. *Small* **2005**, *1*, 632–635. [CrossRef] [PubMed]

205. Kato, N.; Kawashima, T.; Shibata, T.; Mineta, T.; Makino, E. Micromachining of a newly designed AFM probe integrated with hollow microneedle for cellular function analysis. *Microelectron. Eng.* **2010**, *87*, 1185–1189. [CrossRef]

206. Meister, A.; Gabi, M.; Behr, P.; Studer, P.; Vörös, J.; Niedermann, P.; Bitterli, J.; Polesel-Maris, J.; Liley, M.; Heinzelmann, H.; et al. FluidFM: Combining atomic force microscopy and nanofluidics in a universal liquid delivery system for single cell applications and beyond. *Nano Lett.* **2009**, *9*, 2501–2507. [CrossRef] [PubMed]

207. Guillaume-Gentil, O.; Potthoff, E.; Ossola, D.; Franz, C.M.; Zambelli, T.; Vorholt, J.A. Force-controlled manipulation of single cells: From AFM to FluidFM. *Trends Biotechnol.* **2014**, *32*, 381–388. [CrossRef] [PubMed]

208. Dörig, P.; Stiefel, P.; Behr, P.; Sarajlic, E.; Bijl, D.; Gabi, M.; Vörös, J.; Vorholt, J.A.; Zambelli, T. Force-controlled spatial manipulation of viable mammalian cells and micro-organisms by means of FluidFM technology. *Appl. Phys. Lett.* **2010**, *97*, 023701. [CrossRef]

209. Martinez, V.; Forró, C.; Weydert, S.; Aebersold, M.J.; Dermutz, H.; Guillaume-Gentil, O.; Zambelli, T.; Vörös, J.; Demkó, L. Controlled single-cell deposition and patterning by highly flexible hollow cantilevers. *Lab Chip* **2016**, *16*, 1663–1674. [CrossRef] [PubMed]

210. Sanford, K.K.; Earle, W.R.; Likely, G.D. The growth in vitro of single isolated tissue cells. *J. Nat. Cancer Inst.* **1948**, *9*, 229–246. [PubMed]

211. Anis, Y.H.; Holl, M.R.; Meldrum, D.R. Automated selection and placement of single cells using vision-based feedback control. *IEEE Trans. Autom. Sci. Eng.* **2010**, *7*, 598. [CrossRef]

212. Fröhlich, J.; König, H. New techniques for isolation of single prokaryotic cells. *FEMS Microbiol. Rev.* **2000**, *24*, 567–572. [CrossRef] [PubMed]

213. Gregoire, S.; Xiao, J.; Silva, B.B.; Gonzalez, I.; Agidi, P.S.; Klein, M.I.; Ambatipudi, K.S.; Rosalen, P.L.; Bauserman, R.; Waugh, R.E.; et al. Role of glucosyltransferase B in interactions of *Candida albicans* with *Streptococcus mutans* and with an experimental pellicle on hydroxyapatite surfaces. *Appl. Environ. Microbiol.* **2011**, *77*, 6357–6367. [CrossRef] [PubMed]

214. Lu, Z.; Moraes, C.; Ye, G.; Simmons, C.A.; Sun, Y. Single cell deposition and patterning with a robotic system. *PLoS ONE* **2010**, *5*, e13542. [CrossRef] [PubMed]

215. Környei, Z.; Beke, S.; Mihálffy, T.; Jelitai, M.; Kovács, K.J.; Szabó, Z.; Szabó, B. Cell sorting in a Petri dish controlled by computer vision. *Sci. Rep.* **2013**, *3*, 1088. [CrossRef] [PubMed]

216. Ungai-Salánki, R.; Gerecsei, T.; Fürjes, P.; Orgovan, N.; Sándor, N.; Holczer, E.; Horvath, R.; Szabó, B. Automated single cell isolation from suspension with computer vision. *Sci. Rep.* **2016**, *6*, 20375. [CrossRef] [PubMed]

217. Roder, P.; Hille, C. A Multifunctional frontloading approach for repeated recycling of a pressure-controlled AFM micropipette. *PLoS ONE* **2015**, *10*, e0144157. [CrossRef] [PubMed]

218. Lee, S.; Jeong, W.; Beebe, D.J. Microfluidic valve with cored glass microneedle for microinjection. *Lab Chip* **2003**, *3*, 164–167. [CrossRef] [PubMed]

219. Chung, B.G.; Lin, F.; Jeon, N.L. A microfluidic multi-injector for gradient generation. *Lab Chip* **2006**, *6*, 764–768. [CrossRef] [PubMed]

220. Stiefel, P.; Schmidt, F.I.; Dörig, P.; Behr, P.; Zambelli, T.; Vorholt, J.A.; Mercer, J. Cooperative vaccinia infection demonstrated at the single-cell level using FluidFM. *Nano Lett.* **2012**, *12*, 4219–4227. [CrossRef] [PubMed]

221. Ramser, K.; Hanstorp, D. Optical manipulation for single-cell studies. *J. Biophotonics* **2010**, *3*, 187–206. [CrossRef] [PubMed]

222. Ashkin, A. Acceleration and trapping of particles by radiation pressure. *Phys. Rev. Lett.* **1970**, *24*, 156. [CrossRef]

223. Neuman, K.C.; Block, S.M. Optical trapping. *Rev. Sci. Instrum.* **2004**, *75*, 2787–2809. [CrossRef] [PubMed]

224. Ashkin, A.; Dziedzic, J.M.; Bjorkholm, J.E.; Chu, S. Observation of a single-beam gradient force trap for dielectric particles. *Opt. Lett.* **1986**, *11*, 288–290. [CrossRef] [PubMed]

225. Neuman, K.C.; Liou, G.F.; Block, S.M.; Bergman, K. Characterization of photodamage induced by optical tweezers. In *Conference on Lasers and Electro-Optics*; Scifres, D., Weiner, A., Eds.; paper CTuR1; Optical Society of America: Washington, DC, USA, 1998.

226. Maghelli, N.; Tolić-Nørrelykke, I.M. Optical trapping and laser ablation of microtubules in fission yeast. *Methods Cell. Biol.* **2010**, *97*, 173–183. [PubMed]

227. Simmons, R.M.; Finer, J.T.; Chu, S.; Spudich, J.A. Quantitative measurements of force and displacement using an optical trap. *Biophys. J.* **1996**, *70*, 1813–1822. [CrossRef]

228. Grimbergen, J.A.; Visscher, K.; de Gomes Mesquita, D.S.; Brakenhoff, G.J. Isolation of single yeast cells by optical trapping. *Yeast* **1993**, *9*, 723–732. [CrossRef] [PubMed]

229. Arai, F.; Ng, C.; Maruyama, H.; Ichikawa, A.; El-Shimy, H.; Fukuda, T. On chip single-cell separation and immobilization using optical tweezers and thermosensitive hydrogel. *Lab Chip* **2005**, *5*, 1399–1403. [CrossRef] [PubMed]

230. Singh, G.P.; Creely, C.M.; Volpe, G.; Grötsch, H.; Petrov, D. Real-time detection of hyperosmotic stress response in optically trapped single yeast cells using Raman microspectroscopy. *Anal. Chem.* **2005**, *77*, 2564–2568. [CrossRef] [PubMed]

231. Aabo, T.; Banás, A.R.; Glückstad, J.; Siegumfeldt, H.; Arneborg, N. BioPhotonics workstation: A versatile setup for simultaneous optical manipulation, heat stress, and intracellular pH measurements of a live yeast cell. *Rev. Sci. Instrum.* **2011**, *82*, 083707. [CrossRef] [PubMed]

232. Eriksson, E.; Scrimgeour, J.; Graneli, A.; Ramser, K.; Wellander, R.; Enger, J.; Hanstorp, D.; Goksör, M. Optical manipulation and microfluidics for studies of single cell dynamics. *J. Opt. A Pure Appl. Opt.* **2007**, *9*, S113–S121. [CrossRef]

233. Dholakia, K.; Reece, P. Optical micromanipulation takes hold. *Nanotoday* **2006**, *1*, 18–27. [CrossRef]

234. Castelein, M.; Rouxhet, P.G.; Pignon, F.; Magnin, A.; Piau, J.M. Single-cell adhesion probed in-situ using optical tweezers: A case study with *Saccharomyces cerevisiae*. *J. Appl. Phys.* **2012**, *111*, 114701. [CrossRef]

235. Landenberger, B.; Höfemann, H.; Wadle, S.; Rohrbach, A. Microfluidic sorting of arbitrary cells with dynamic optical tweezers. *Lab Chip* **2012**, *12*, 3177–3183. [CrossRef] [PubMed]

236. Werner, M.; Merenda, F.; Piguet, J.; Salathé, R.P.; Vogel, H. Microfluidic array cytometer based on refractive optical tweezers for parallel trapping, imaging and sorting of individual cells. *Lab Chip* **2011**, *11*, 2432–2439. [CrossRef] [PubMed]

237. Tam, J.M.; Castro, C.E.; Heath, R.J.; Cardenas, M.L.; Xavier, R.J.; Lang, M.J.; Vyas, J.M. Control and manipulation of pathogens with an optical trap for live cell imaging of intercellular interactions. *PLoS ONE* **2010**, *5*, e15215. [CrossRef] [PubMed]

238. Tam, J.M.; Castro, C.E.; Heath, R.J.; Mansour, M.K.; Cardenas, M.L.; Xavier, R.J.; Lang, M.J.; Vyas, J.M. Use of an optical trap for study of host-pathogen interactions for dynamic live cell imaging. *J. Vis. Exp.* **2011**, *53*, 3123. [CrossRef] [PubMed]

239. Eriksson, E.; Sott, K.; Lundqvist, F.; Sveningsson, M.; Scrimgeour, J.; Hanstorp, D.; Goksör, M.; Granéli, A. A microfluidic device for reversible environmental changes around single cells using optical tweezers for cell selection and positioning. *Lab Chip* **2010**, *10*, 617–625. [CrossRef] [PubMed]

240. Arneborg, N.; Siegumfeldt, H.; Andersen, G.H.; Nissen, P.; Daria, V.R.; Rodrigo, P.J.; Glückstad, J. Interactive optical trapping shows that confinement is a determinant of growth in a mixed yeast culture. *FEMS Microbiol. Lett.* **2005**, *245*, 155–159. [CrossRef] [PubMed]

241. Charrunchon, S.; Limtrakul, J.; Chattham, N. Growth pattern of yeast cells studied under optical tweezers. In *Frontiers in Optics*; 2012/Laser Science XXVIII, OSA Technical Digest (Online); Paper FW1G.6; Optical Society of America: Rochester, NY, USA, 2012.

242. Ando, J.; Bautista, G.; Smith, N.; Fujita, K.; Daria, V.R. Optical trapping and surgery of living yeast cells using a single laser. *Rev. Sci. Instrum.* **2008**, *79*, 103705. [CrossRef] [PubMed]

243. Hu, S.; Sun, D. Automated transportation of single cells using robot-tweezer manipulation system. *J. Lab. Autom.* **2011**, *16*, 263–270. [CrossRef] [PubMed]

244. Gong, Y.; Huang, W.; Liu, Q.F.; Wu, Y.; Rao, Y.; Peng, G.D.; Lang, J.; Zhang, K. Graded-index optical fiber tweezers with long manipulation length. *Opt. Express* **2014**, *22*, 25267–25276. [CrossRef] [PubMed]

245. Gustavsson, A.K.; van Niekerk, D.D.; Adiels, C.B.; du Preez, F.B.; Goksör, M.; Snoep, J.L. Sustained glycolytic oscillations in individual isolated yeast cells. *FEBS J.* **2012**, *279*, 2837–2847. [CrossRef] [PubMed]

246. Gustavsson, A.K.; van Niekerk, D.D.; Adiels, C.B.; Kooi, B.; Goksör, M.; Snoep, J.L. Allosteric regulation of phosphofructokinase controls the emergence of glycolytic oscillations in isolated yeast cells. *FEBS J.* **2014**, *281*, 2784–2793. [CrossRef] [PubMed]

247. Habaza, M.; Gilboa, B.; Roichman, Y.; Shaked, N.T. Tomographic phase microscopy with 180° rotation of live cells in suspension by holographic optical tweezers. *Opt. Lett.* **2015**, *40*, 1881–1884. [CrossRef] [PubMed]

248. Jing, P.; Wu, J.; Liu, G.W.; Keeler, E.G.; Pun, S.H.; Lin, L.Y. Photonic crystal optical tweezers with high efficiency for live biological samples and viability characterization. *Sci. Rep.* **2016**, *6*, 19924. [CrossRef] [PubMed]

249. Tolić-Nørrelykke, I.M.; Munteanu, E.L.; Thon, G.; Oddershede, L.; Berg-Sørensen, K. Anomalous diffusion in living yeast cells. *Phys. Rev. Lett.* **2004**, *93*, 078102. [CrossRef] [PubMed]

250. Sacconi, L.; Tolić-Nørrelykke, I.M.; Stringari, C.; Antolini, R.; Pavone, F.S. Optical micromanipulations inside yeast cells. *Appl. Opt.* **2005**, *44*, 2001–2007. [CrossRef] [PubMed]

251. Tolic-Nørrelykke, I.M.; Sacconi, L.; Stringari, C.; Raabe, I.; Pavone, F.S. Nuclear and division-plane positioning revealed by optical micromanipulation. *Curr. Biol.* **2005**, *15*, 1212–1216. [CrossRef] [PubMed]

252. Jeon, J.H.; Tejedor, V.; Burov, S.; Barkai, E.; Selhuber-Unkel, C.; Berg-Sørensen, K.; Oddershede, L.; Metzler, R. In vivo anomalous diffusion and weak ergodicity breaking of lipid granules. *Phys. Rev. Lett.* **2011**, *106*, 048103. [CrossRef] [PubMed]

253. Mas, J.; Richardson, A.C.; Reihani, S.N.; Oddershede, L.B.; Berg-Sørensen, K. Quantitative determination of optical trapping strength and viscoelastic moduli inside living cells. *Phys. Biol.* **2013**, *10*, 046006. [CrossRef] [PubMed]

254. Difato, F.; Pinato, G.; Cojoc, D. Cell signalling periments driven by optical manipulation. *Int. J. Mol. Sci.* **2013**, *14*, 8963–8984. [CrossRef] [PubMed]

255. Kotsifaki, D.G.; Makropoulou, M.; Serafetinides, A. Near infrared optical tweezers and nanosecond ablation on yeast and algae cells. In Proceedings of the SPIE 17th International School on Quantum Electronics: Laser Physics and Applications, Nessebar, Bulgaria, 24–28 September 2012; Volume 8770.

256. Oddershede, L.B. Force probing of individual molecules inside the living cell is now a reality. *Nat. Chem. Biol.* **2012**, *8*, 879–886. [CrossRef] [PubMed]

257. Norregaard, K.; Jauffred, L.; Berg-Sørensen, K.; Oddershede, L.B. Optical manipulation of single molecules in the living cell. *Phys. Chem. Chem. Phys.* **2014**, *16*, 12614–12624. [CrossRef] [PubMed]

258. Toriello, N.M.; Douglas, E.S.; Mathies, R.A. Microfluidic device for electric field-driven single-cell capture and activation. *Anal. Chem.* **2005**, *77*, 6935–6941. [CrossRef] [PubMed]

259. Koyama, S.; Tsubouchi, T.; Usui, K.; Uematsu, K.; Tame, A.; Nogi, Y.; Ohta, Y.; Hatada, Y.; Kato, C.; Miwa, T.; et al. Involvement of flocculin in negative potential-applied ITO electrode adhesion of yeast cells. *FEMS Yeast Res.* **2015**, *15*. [CrossRef] [PubMed]

260. Qian, C.; Huang, H.; Chen, L.; Li, X.; Ge, Z.; Chen, T.; Yang, Z.; Sun, L. Dielectrophoresis for bioparticle manipulation. *Int. J. Mol. Sci.* **2014**, *15*, 18281–18309. [CrossRef] [PubMed]

261. Norde, W. *Colloids and Interfaces in Life Sciences*; Marcel Dekker: Monticello, NY, USA, 2003.

262. Jubery, T.Z.; Srivastava, S.K.; Dutta, P. Dielectrophoretic separation of bioparticles in microdevices: A review. *Electrophoresis* **2014**, *35*, 691–713. [CrossRef] [PubMed]

263. Kang, K.H.; Xuan, X.C.; Kang, Y.J.; Li, D.Q. Effects of dc-dielectrophoretic force on particle trajectories in microchannels. *J. Appl. Phys.* **2006**, *99*, 064702. [CrossRef]

264. Khoshmanesh, K.; Nahavandi, S.; Baratchi, S.; Mitchell, A.; Kalantar-zadeh, K. Dielectrophoretic platforms for bio-microfluidic systems. *Biosens. Bioelectron.* **2011**, *26*, 1800–1814. [CrossRef] [PubMed]

265. Koklu, M.; Park, S.; Pillai, S.D.; Beskok, A. Negative dielectrophoretic capture of bacterial spores in food matrices. *Biomicrofluidics* **2010**, *4*. [CrossRef] [PubMed]

266. Cheng, I.F.; Froude, V.E.; Zhu, Y.; Chang, H.C.; Chang, H.C. A continuous high-throughput bioparticle sorter based on 3D traveling-wave dielectrophoresis. *Lab Chip* **2009**, *9*, 3193–3201. [CrossRef] [PubMed]

267. Hunt, T.P.; Westervelt, R.M. Dielectrophoresis tweezers for single cell manipulation. *Biomed. Microdevices* **2006**, *8*, 227–230. [CrossRef] [PubMed]

268. Cetin, B.; Kang, Y.; Wu, Z.; Li, D. Continuous particle separation by size via AC-dielectrophoresis using a lab-on-a-chip device with 3-D electrodes. *Electrophoresis* **2009**, *30*, 766–772. [CrossRef] [PubMed]

269. Docoslis, A.; Kalogerakis, N.; Behie, L.A. Dielectrophoretic forces can be safely used to retain viable cells in perfusion cultures of animal cells. *Cytotechnology* **1999**, *30*, 133–142. [CrossRef] [PubMed]

270. Pethig, R.; Menachery, A.; Pells, S.; de Sousa, P. Dielectrophoresis: A review of applications for stem cell research. *J. Biomed. Biotechnol.* **2010**, *2010*, 182581. [CrossRef] [PubMed]

271. Li, S.; Li, M.; Bougot-Robin, K.; Cao, W.; Chau, I.Y.Y.; Li, W.; Wen, W. High-throughput particle manipulation by hydrodynamic, electrokinetic, and dielectrophoretic effects in an integrated microfluidic chip. *Biomicrofluidics* **2013**, *7*, 24106. [CrossRef] [PubMed]

272. Li, P.C.; Harrison, D.J. Transport, manipulation, and reaction of biological cells on-chip using electrokinetic effects. *Anal. Chem.* **1997**, *69*, 1564–1568. [CrossRef] [PubMed]
273. Kodama, T.; Osaki, T.; Kawano, R.; Kamiya, K.; Miki, N.; Takeuchi, S. Round-tip dielectrophoresis-based tweezers for single micro-object manipulation. *Biosens. Bioelectron.* **2013**, *47*, 206–212. [CrossRef] [PubMed]
274. Rosenthal, A.; Voldman, J. Dielectrophoretic traps for single-particle patterning. *Biophys. J.* **2005**, *88*, 2193–2205. [CrossRef] [PubMed]
275. Lee, H.; Hunt, T.P.; Westervelt, R.M. Magnetic and electric manipulation of a single cell in fluid. *Mater. Res. Symp. Proc.* **2004**, *820*. [CrossRef]
276. Lee, H.; Purdon, A.M.; Westervelt, R.M. Micromanipulation of biological systems with microelectromagnets. *IEEE Trans. Magn.* **2004**, *40*, 2991–2993. [CrossRef]
277. Jang, L.S.; Huang, P.H.; Lan, K.C. Single-cell trapping utilizing negative dielectrophoretic quadrupole and microwell electrodes. *Biosens. Bioelectron.* **2009**, *24*, 3637–3644. [CrossRef] [PubMed]
278. Arnold, W.M. Positioning and levitation media for the separation of biological cells. *Ind. Appl. IEEE Trans.* **2001**, *37*, 1468–1475. [CrossRef]
279. Gagnon, Z.; Mazur, J.; Chang, H.C. Glutaraldehyde enhanced dielectrophoretic yeast cell separation. *Biomicrofluidics* **2009**, *3*, 44108. [CrossRef] [PubMed]
280. Fatoyinbo, H.O.; Kamchis, D.; Whattingham, R.; Ogin, S.L.; Hughes, M.P. A high-throughput 3-D composite dielectrophoretic separator. *IEEE Trans. Biomed. Eng.* **2005**, *52*, 1347–1349. [CrossRef] [PubMed]
281. Urdaneta, M.; Smela, E. Multiple frequency dielectrophoresis. *Electrophoresis* **2007**, *28*, 3145–3155. [CrossRef] [PubMed]
282. Salomon, S.; Leichlé, T.; Nicu, L. A dielectrophoretic continuous flow sorter using integrated microelectrodes coupled to a channel constriction. *Electrophoresis* **2011**, *32*, 1508–1514. [CrossRef] [PubMed]
283. Huang, Y.; Hölzel, R.; Pethig, R.; Wang, X.B. Differences in the AC electrodynamics of viable and non-viable yeast cells determined through combined dielectrophoresis and electrorotation studies. *Phys. Med. Biol.* **1992**, *37*, 1499–1517. [CrossRef] [PubMed]
284. Hölzel, R. Electrorotation of single yeast cells at frequencies between 100 Hz and 1.6 GHz. *Biophys. J.* **1997**, *73*, 1103–1109. [CrossRef]
285. Fatoyinbo, H.O.; Hoettges, K.F.; Hughes, M.P. Rapid-on-chip determination of dielectric properties of biological cells using imaging techniques in a dielectrophoresis dot microsystem. *Electrophoresis* **2008**, *29*, 3–10. [CrossRef] [PubMed]
286. Soffe, R.; Tang, S.Y.; Baratchi, S.; Nahavandi, S.; Nasabi, M.; Cooper, J.M.; Mitchell, A.; Khoshmanesh, K. Controlled rotation and vibration of patterned cell clusters using dielectrophoresis. *Anal. Chem.* **2015**, *87*, 2389–2395. [CrossRef] [PubMed]
287. Hunt, T.P.; Issadore, D.; Westervelt, R.M. Integrated circuit/microfluidic chip to programmably trap and move cells and droplets with dielectrophoresis. *Lab. Chip* **2008**, *8*, 81–87. [CrossRef] [PubMed]
288. Witek, M.A.; Wei, S.; Vaidya, B.; Adams, A.A.; Zhu, L.; Stryjewski, W.; McCarley, R.L.; Soper, S.A. Cell transport via electromigration in polymer-based microfluidic devices. *Lab Chip* **2004**, *4*, 464–472. [CrossRef] [PubMed]
289. Frenea-Robin, M.; Chetouani, H.; Haddour, N.; Rostaing, H.; Laforet, J.; Reyne, G. Contactless diamagnetic trapping of living cells onto a micromagnet array. *Conf. Proc. IEEE Eng. Med. Biol. Soc.* **2008**, *2008*, 3360–3363. [PubMed]
290. Yasukawa, T.; Nagamine, K.; Horiguchi, Y.; Shiku, H.; Koide, M.; Itayama, T.; Shiraishi, F.; Matsue, T. Electrophoretic cell manipulation and electrochemical gene-function analysis based on a yeast two-hybrid system in a microfluidic device. *Anal. Chem.* **2008**, *80*, 3722–3727. [CrossRef] [PubMed]
291. Moncada-Hernandez, H.; Baylon-Cardiel, J.L.; Pérez-González, V.H.; Lapizco-Encinas, B.H. Insulator-based dielectrophoresis of microorganisms: Theoretical and experimental results. *Electrophoresis* **2011**, *32*, 2502–2511. [CrossRef] [PubMed]
292. Yue, T.; Nakajima, M.; Tajima, H.; Fukuda, T. Fabrication of microstructures embedding controllable particles inside dielectrophoretic microfluidic devices. *Int. J. Adv. Robot. Syst.* **2013**, *10*, 132. [CrossRef]
293. Shi, X.; Shi, Z.; Wang, D.; Ullah, M.W.; Yang, G. Microbial cells with a Fe_3O_4 doped hydrogel extracellular matrix: Manipulation of living cells by magnetic stimulus. *Macromol. Biosci.* **2016**. [CrossRef] [PubMed]
294. Winkleman, A.; Gudiksen, K.L.; Ryan, D.; Whitesides, G.M.; Greenfield, D.; Prentiss, M. A magnetic trap for living cells suspended in a paramagnetic buffer. *Appl. Phys. Lett.* **2004**, *85*, 2411. [CrossRef]

295. Neuman, K.C.; Nagy, A. Single-molecule force spectroscopy: Optical tweezers, magnetic tweezers and atomic force microscopy. *Nat. Methods* **2008**, *5*, 491–505. [CrossRef] [PubMed]

296. Fisher, J.K.; Cribb, J.; Desai, K.V.; Vicci, L.; Wilde, B.; Keller, K.; Taylor, R.M.; Haase, J.; Bloom, K.; O'Brien, E.T.; et al. Thin-foil magnetic force system for high-numerical-aperture microscopy. *Rev. Sci. Instrum.* **2006**, *77*. [CrossRef] [PubMed]

297. Yan, J.; Skoko, D.; Marko, J.F. Near-field-magnetic-tweezer manipulation of single DNA molecules. *Phys. Rev. E Stat. Nonlinear Soft Matter Phys.* **2004**, *70*, 011905. [CrossRef] [PubMed]

298. Chacko, J.V.; Zanacchi, F.C.; Diaspro, A. Probing cytoskeletal structures by coupling optical superresolution and AFM techniques for a correlative approach. *Cytoskeleton (Hoboken)* **2013**, *70*, 729–740. [CrossRef] [PubMed]

299. Chacko, J.V.; Harke, B.; Canale, C.; Diaspro, A. Cellular level nanomanipulation using atomic force microscope aided with superresolution imaging. *J. Biomed. Opt.* **2014**, *19*, 105003. [CrossRef] [PubMed]

300. Asakawa, H.; Hiraoka, Y.; Haraguchi, T. A method of correlative light and electron microscopy for yeast cells. *Micron* **2014**, *61*, 53–61. [CrossRef] [PubMed]

301. Kobayashi, S.; Iwamoto, M.; Haraguchi, T. Live correlative light-electron microscopy to observe molecular dynamics in high resolution. *Microscopy (Oxf.)* **2016**, *65*, 296–308. [CrossRef] [PubMed]

302. Wolff, G.; Hagen, C.; Grünewald, K.; Kaufmann, R. Towards correlative super-resolution fluorescence and electron cryo-microscopy. *Biol. Cell* **2016**, *108*, 245–258. [CrossRef] [PubMed]

303. Heinisch, J.J.; Lipke, P.N.; Beaussart, A.; El Kirat Chatel, S.; Dupres, V.; Alsteens, D.; Dufrêne, Y.F. Atomic force microscopy—Looking at mechanosensors on the cell surface. *J. Cell Sci.* **2012**, *125*, 4189–4195. [CrossRef] [PubMed]

304. Conroy, R. Force spectroscopy with optical and magnetic tweezers. In *Handbook of Molecular Force Spectroscopy*; Noy, A., Ed.; Springer: New York, NY, USA, 2008; pp. 23–96.

305. Alonso-Sarduy, L.; de Los Rios, P.; Benedetti, F.; Vobornik, D.; Dietler, G.; Kasas, S.; Longo, G. Real-time monitoring of protein conformational changes using a nano-mechanical sensor. *PLoS ONE* **2014**, *9*, e103674. [CrossRef] [PubMed]

fermentation

MDPI

Review

Starter Cultures for Sparkling Wine

Carmela Garofalo [1], Mattia Pia Arena [1], Barbara Laddomada [2], Maria Stella Cappello [2], Gianluca Bleve [2], Francesco Grieco [2], Luciano Beneduce [1], Carmen Berbegal [1], Giuseppe Spano [1,*] and Vittorio Capozzi [1]

[1] Dipartimento di Scienze Agrarie, degli Alimenti e dell'Ambiente, Università di Foggia, via Napoli 25, Foggia 71100, Italy; carmela.garofalo@unifg.it (C.G.); mattiapia.arena@unifg.it (M.P.A.); luciano.beneduce@unifg.it (L.B.); carmen.berbegal@unifg.it (C.B.); vittorio.capozzi@unifg.it (V.C.)

[2] Istituto di Scienze delle Produzioni Alimentari, Consiglio Nazionale delle Ricerche, Unità Operativa di Supporto di Lecce, Lecce 73100, Italy; barbara.laddomada@ispa.cnr.it (B.L.); maristella.cappello@ispa.cnr.it (M.S.C.); gianluca.bleve@ispa.cnr.it (G.B.); francesco.grieco@ispa.cnr.it (F.G.)

* Correspondence: giuseppe.spano@unifg.it; Tel.: +39-881-589303

Academic Editor: Ronnie G. Willaert
Received: 26 October 2016; Accepted: 8 December 2016; Published: 14 December 2016

Abstract: The sparkling wine market has expanded in recent years, boosted by the increasing demand of the global market. As for other fermented beverages, technological yeasts and bacteria selected to design commercial starter cultures represent key levers to maximize product quality and safety. The increasing economic interest in the sector of sparkling wine has also implied a renewed interest in microbial resource management. In this review, after a brief introduction, we report an overview of the main characterization criteria in order to select *Saccharomyces cerevisiae* strains suitable for use as starter cultures for the production of base wines and to drive re-fermentation of base wines to obtain sparkling wines. Particular attention has been reserved to the technological characterization aspects of re-fermenting phenotypes. We also analysed the possible uses of selected non-*Saccharomyces* and malolactic strains in order to differentiate specific productions. Finally, we highlighted the main safety aspects related to microbes of enological interest and underlined some microbial-based biotechnological applications helpful to pursue product and process innovations. Overall, the sparkling wine industry may find a relevant benefit from the exploitation of the wide resources associated with vineyard/wine microbial diversity.

Keywords: sparkling wine; starter cultures; *Saccharomyces cerevisiae*; non-*Saccharomyces* autolysis; flocculation; alcoholic fermentation; re-fermentation

1. Introduction

Humans have produced alcoholic beverages for millennia and these products have been traditionally used for medicinal, nutritional, and social purposes [1,2]. During the centuries, the technical procedures for their production have continuously evolved, since the discovery of spontaneous fermentations to the industrial application of starter cultures. The microbial strains mainly used for this last purpose belong to the *Saccharomyces* species. However, non-*Saccharomyces* species, previously considered spoilage yeasts, have also been recently used as fermentation starters, in the perspective of wines designed to respond to consumer demands [3–6].

During alcoholic fermentation, yeasts produce several compounds, mainly ethanol and carbon dioxide, with the latter released directly into the atmosphere if the process is conducted in an open vessel. This is what happens during the production of most wines, normally defined "still" wines for the negligible amounts of carbon dioxide that they still contain. Contrariwise, wines containing a relevant concentration of carbon dioxide are referred as "effervescent" wines, distinguished into semi-sparkling (1–2.5 atmospheres of pressure in the bottle) and sparkling varieties (3–6 atmospheres of pressure in the bottle) [7].

Almost all sparkling wines are the result of two fermentation steps. During the first fermentation, the must is converted into wine (usually denoted as "base" wine), whereas in the second fermentation step, the base wine is firstly added with several ingredients (e.g., sucrose, yeasts cells, nitrogen source) and then re-fermented in a cellar for, at least, 9–12 months [8].

The sparkling wine market has expanded in recent years, boosted by a high global consumer demand [9,10]. The production of sparkling wine has significantly increased, showing a rise of 40% in the last ten years, while that of non-sparkling wines only increased by 7% over the same period [11].

Sparkling wines have an important economic impact due to their high added value. Moreover, considering the consumers' attention toward quality and safety of fermented beverages, research in the sparkling wine sector is nowadays also focuses on biotechnological innovations to improve product qualities, to simplify the production process and, at the same time, to enhance the preservation of typical and unique product characteristics [12–14].

Among the steps of production of sparkling wine, an important phase for ensuring the quality of the final product is the aging. In fact, at the end of second fermentation, sparkling wines undergo an aging period during which yeast autolysis occurs, with the consequent release of several cellular compounds, such as amino acids, proteins, carbohydrates, and lipids, that improve the quality of sparkling wines [15].

Modern biotechnologies can be used not only to improve the quality of sparkling wines, but also to reduce their production time and costs [8,14,16,17]; for example, some authors suggest new methods to accelerate the above-described autolysis process. Starter cultures for sparkling wine production need to be selected in order to produce either quality base wine or to vigorously promote the second fermentation, which occurs in a harsh environment, mainly due to the elevated ethanol content, low pH (2.8–3.3), and carbon dioxide-induced pressure [18].

2. Sparkling Wine: Production Process, Legislation, and Classification

Sparkling wines can be produced according to two main procedures, the traditional, also called as "champenoise", and the Charmat methods. The traditional method, is performed by an in-bottle secondary fermentation, while in the Charmat method, secondary fermentation is carried out in hermetically-sealed tanks [11].

2.1. Production of Sparkling Wine Using the Charmat Method

The Charmat method is characterized by sealed tanks provided with agitating mechanisms, with the aim to mix the yeast uniformly into the base wine during secondary fermentation [8]. Base wines are usually corrected with 20–24 g/L of sugar and then pasteurized with the aim to accelerate sucrose hydrolysis. Briefly, the base wine integrated with sugar is added with yeast into a pressure tank made of stainless steel, built to resist the pressure. When all of the sugar is transformed into alcohol and carbon dioxide, the yeasts are removed and the wine is bottled in an isobaric, refrigerated environment. The duration of fermentation usually influences the quality of the final product and a prolonged fermentation protects the wine aroma and allows maintaining bubbles that are more durable. At the end of secondary fermentation and after clarification, sparkling wine is bottled. After that, it should be aged at least 20 days before sale, during which aging wine remains in contact with yeast lees [19]. The Charmat method is simpler and cheaper than the traditional one. However, the process cannot be used for sparkling wines with specific regional designations.

2.2. Production of Sparkling Wine Using the Traditional Method

The traditional method is referred to as "méthode champenoise", but this expression can be officially used only for sparkling wines produced in the Champagne region (EU regulation number 3309/85). Consequently, all other sparkling wines can be identified by the expression "traditional method", "classic method", or similar terms.

Production of sparkling wine using traditional method includes two steps (Figure 1): primary fermentation and secondary fermentation. During primary fermentation, the grape juice is converted into base wine, while during secondary fermentation (also known as *"prise de mousse"*) the alcoholic fermentation of the sucrose-base-wine mixture produces the sparkling wine, with its typical characteristics, flavour, and foam.

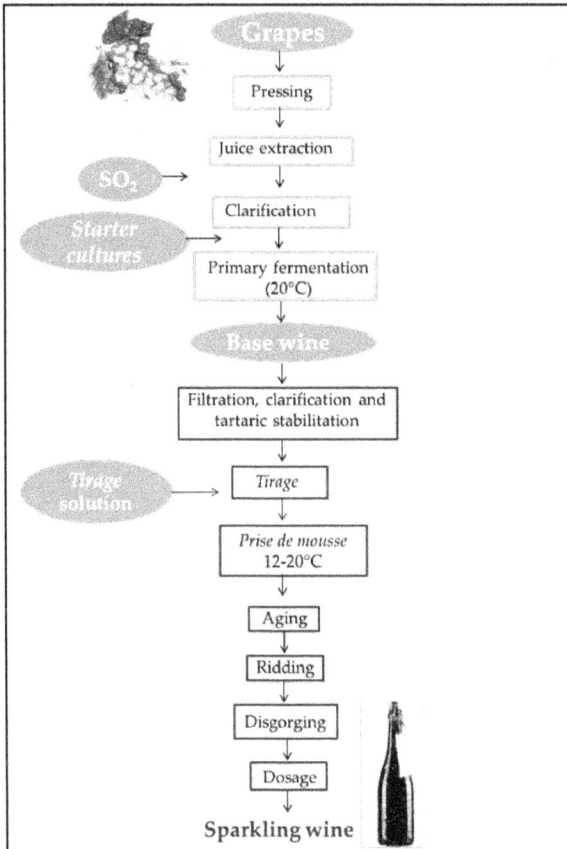

Figure 1. Flowchart for sparkling wine production using the traditional, or *champenoise*, method.

A typical base wine usually presents a moderate alcohol concentration (10%–11% *v/v*), low levels of sugar and acetic acid (volatile acidity), and a high concentration of other organic acids (total acidity) [8]. In particular, a significant value of the total acidity of base wine (about 12–18 g/L measured as tartaric acid), represents a critical point for sparkling wine production, because during the production process, the total acidity may decrease due to several factors, such as malic acid degradation by yeasts and lactic acid bacteria and potassium bitartrate precipitations [20].

A common practice to avoid insufficient total acidity is to slightly anticipate harvesting with respect to traditional white wine grapes. Harvesting, in fact, is a very delicate step during base wine production. Grapes are usually hand-picked and collected into small tanks to avoid berry breakage. Only sound grapes are collected and quickly transported to the cellar, in order to prevent spontaneous fermentation. Usually, grape berries are immediately pressed, without crushing to avoid oxidations, macerations, and the development of flat aromas and browning, to obtain a good quality must. Grape juice extraction must be extremely meticulous to avoid vegetal and bitter defects, due to an excessive maceration of skins. Base wines should also have a fruity flavor and a pale color. Indeed, color extraction should be avoided, especially for white sparkling wines obtained from red grapes, also named *"blanc de noirs"* [21]. After must extraction, the next steps are (i) the addition of sulphur dioxide (apart from that added before crushing) to prevent oxidation and undesired fermentation; (ii) the clarification with pectolytic enzymes to remove solids and minimize phenol oxidations; (iii) the inoculation of starter cultures of *Saccharomyces cerevisiae*, which drive the alcoholic fermentation process in stainless steel tanks under temperature control.

At the end of primary fermentation base wines are subjected to several manipulations, such as further clarification, decantation, and filtration. Common practices also include the addition of bentonite and cold precipitation, to promote protein precipitation and potassium bitartrate precipitation, respectively. After such treatments, the base wine is ready for the secondary fermentation and selected starter cultures (*S. cerevisiae*) are added within a so-called *"tirage"* solution (containing ingredients such as saccharose 20–25 g/L, yeasts, grape must or wine, and bentonite) [22]. After the addition of the *tirage* solution, the wine is bottled and the bottles are sealed with a crown cap, that underneath it has a *"bidule"*, i.e., a plastic cylinder where the lees will accumulate. The bottles are then horizontally stacked in special aging rooms at low temperature (12–15 °C). Sparkling aging usually takes place horizontally, as this position promotes an efficient contact between wine and yeast sediment, with a slow release of several compounds originating both from yeasts and wine. After the production of the desired CO_2 concentration and aging, a disgorging procedure, traditionally denoted as "dégorgement" [22], carries out the removal of lees.

The kinetics of the secondary fermentation depend on various factors, such as the yeast species/strains, temperature, and chemical composition of base wine. Usually, at 12–15 °C, the secondary fermentation takes almost 15–45 days and can be monitored by checking the internal pressure using an afrometer.

Aging duration is regulated by national legislation; as a consequence, it may vary according to country. Nevertheless, sparkling wine maturation is a slow process and it takes from a minimum of nine months for the "Cava" (Spain) to 12 months for "Talento" (Italy) or "Champagne" (France) wines [23].

A prolonged aging is essential to improve and develop the organoleptic properties of sparkling wine since it is correlated with roundness, flavor, complexity, and foaming [14,15,24]. During aging, the characteristics of sparkling wines change due to the release of yeast cytoplasmic and cell wall compounds into the wine, by the autolysis process promoted by the activity of hydrolytic enzymes. Several authors, who suggested its positive effect on sparkling wine quality, have investigated this biological event [14,24]. Proteins released by yeast into sparkling wine show a positive correlation on "body sensation" and foam stability, while polysaccharides should improve wine stability against protein haze [14]. Contrariwise, yeasts wall should adsorb volatile compounds affecting the aroma of sparkling wine [25].

When the aging is complete, the next step is the riddling or *"remuage"*, i.e., a kind shaking of the bottles to convey yeast lees into the *bidule*. This step is improved by adding a small amount of bentonite to the *tirage* solution and by yeast's aptitude to flocculate. During riddling, each bottle should be hand-rotated one eighth (of the total rotation) each day for 15 days until bottles are practically perpendicular to the floor. Riddling should promote the subsequent disgorging process, during which, lees collected at the neck of the bottle are removed, thanks to freezing and internal pressure in the bottle. Then, it is a common practice to add a dosage solution, traditionally called *"liqueur d'expédition"*, to compensate liquid lost during disgorging. It consists of a mixture of variable composition with pure sparkling wine, sparkling wine containing sucrose, grape must, brandy, SO_2, or other components typical of a determined production area. The dosage solution and its composition influence sparkling wine characteristics, and will give to each sparkling wine a distinctive structure and aroma.

3. Yeast Characterization for Wine Base Production

Wine organoleptic properties are strictly correlated with the physiological and metabolic characteristics of *S. cerevisiae* and non-*Saccharomyces* strains used as starters. Indeed, such microbial component influences the production of several compounds and transform grape compounds with positive or negative effects on fermentative or secondary aromas [26,27].

A starter culture is a microbial strain that is characterized and selected for its fermentation properties. Specific criteria have been indicated to select *S. cerevisiae* starter cultures with exquisite oenological properties. However, the first fundamental step for the selection of oenological starters is the availability of genetics and molecular diagnostic tools that allow a quick and accurate yeast identification, at either species or strain level, and their monitoring during wine fermentation [28–32].

3.1. Yeast Genotypic Characterization: Methods to Differentiate Saccharomyces Cerevisiae Strains

Several molecular techniques have been developed and successfully applied to the identification and characterization of yeasts that allow to differentiate *S. cerevisiae* at the strain level [28,33–38] (Table 1).

Table 1. *Saccharomyces cerevisiae* genotypic characterization.

Molecular Method	Reference
Random amplified polymorphic DNA (RAPD) PCR	[39]
Interdelta sequences analysis	[40]
Pulse field electrophoresis (PFGE) electrophoretic karyotypes	[38]
Mitochondrial DNA (mtDNA) restriction analysis	[33,34,36,41]
Polymorphic microsatellite loci (SSRs, simple sequence repeats)	[40,42]
Multilocus sequence typing (MLST)	[43,44]

The first technique used to reveal *Saccharomyces* strain diversity is pulsed field electrophoresis (PFGE) i.e., separation of intact chromosomes by pulsed field agarose gel electrophoresis, also called electrophoretic karyotyping [38].

Random amplified polymorphic DNA (RAPD-PCR) also has been powerful to differentiate *S. cerevisiae* strains; nevertheless, other methods are more discriminating [41].

Some authors have suggested that mitochondrial DNA (mtDNA) restriction analysis (mtDNA RFLP) could be an efficient technique to differentiate at the strain level [33,34,36,41]. In particular, this molecular technique has been used to check the dominance of *S. cerevisiae* starter cultures, thanks to the marked mtDNA polymorphism of wine *Saccharomyces* strains [45–47].

Another commonly used molecular approach relies on sequencing the interdelta element, whose amplification by PCR allows differentiating at the strain level *S. cerevisiae* strains [33,48]. Other powerful molecular tools for *S. cerevisiae* strain differentiation are the amplification of polymorphic microsatellite loci, also called simple sequence repeats (SSRs) [40,42], the multilocus sequence typing (MLST) and yeast killer virus (virus dsRNA) [43,44,49,50].

3.2. Yeast Technological and Qualitative Characterization for Starter Culture Production

Yeasts, mainly *S. cerevisiae* strains, have a fundamental role during winemaking and alcoholic fermentation. Grape sugars, in particular hexoses, must be rapidly and completely converted into ethanol and CO_2, with the associated production, by the yeasts, of several metabolites important to confer wine typical organoleptic properties (but also the possible release of off-flavors) [51–54]. Nevertheless, it is important to underline that a complete transformation of sugar occurs on dry wine but not on sweet wine.

Must/wine system represents a hostile environment due to several factors, such as high sugar concentration (average 200 g/L), growing ethanol and glycerol amount, low pH (3–3.5), the presence of sulphites, and progressive consumption of nutrients (such as nitrogen sources, vitamins, and lipids) [55].

Usually, starter cultures or autochthonous strains should be selected on the basis of typical oenological traits due to the peculiar characteristics of grape juice, base wine, and to desirable qualities of wines. Indeed, an efficient procedure to characterize *S. cerevisiae* starter strains selected from natural fermenting needs biotechnological tools/criteria to optimize global wine quality [51–55]. Several authors proposed technological and qualitative criteria to select yeast strains with desirable features. Among these, tolerance to alcohol, resistance to sulphur dioxide, several enzymatic activities, osmotic properties, killer factor, and low production of H_2S are determinants [26]. Table 2 reports the most important technological and qualitative criteria to select yeast starter cultures.

Table 2. Yeast technological and qualitative characteristics for starter cultures production.

Technological and Qualitative Characteristics	Reference
Resistance to low pH, sugars, ethanol, and sulphur dioxide contents	[26,46,51,55–57]
Low volatile acidity production	[52,58–60]
Low production of sulphur compounds (H_2S, SO_2)	[26,46,47,51–53,61]
Fermentation vigour	[51]
Desired enzymatic activities (e.g., β-glucosidase, β-xylosidase, protease, polygalacturonase, pectinase, glucanase, xylanase, and decarboxylase activities)	[46,47,57,62,63]
Desired fermentation-associated metabolites (glycerol, succinic acid, acetic acid, acetaldehyde, n-propanol, iso-butanol, isoamyl alcohol, and β-phenylethanol)	[46,47,57,62–64]
Implantation aptitude	[65–67]

Other relevant features include the strain-specific formation of fermentation-associated metabolites (such as glycerol, succinic acid, acetaldehyde, n-propanol, isobutanol, isoamyl alcohol, and β-phenylethanol) and the presence of specific extracellular enzymatic activities (β-glucosidase, β-xylosidase, protease, polygalacturonase, pectinase, glucanase, xylanase, and decarboxylase) [46,47,57,62–64]. Obviously, implantation aptitude of starter cultures also is a criterion to be checked during starter technological selection programs. Indeed, several studies suggested that starter cultures dominance is not always guaranteed, as a function of the diversity associated with the naturally present microbial consortia, and that during winemaking indigenous yeasts can survive and grow, affecting starter dominance [65–67].

4. Yeast Technological Characterization for Secondary Fermentation of Sparkling Wine Production

The starter cultures used for the secondary fermentation in the traditional method need to possess several additional technological properties to those of the yeast used in the primary fermentation (Figure 2).

Chemical composition of the base wine and the sparkling wine production process represent a hostile environment for yeast growth and fermentation efficiency [23]. Base wine usually is characterized by consistent ethanol concentration (about 10%–12% v/v), low pH (2.8–3.5), high total acidity (5–7 g/L H_2SO_4), and total SO_2 contents (50–80 mg/L). In addition to these critical factors, we have to consider low temperatures occurring during the secondary fermentation (10–15 °C) and the high amount of CO_2/high pressure (usually 6 atm) associated with this process [68]. Hence, yeast starter cultures for secondary fermentation have to be selected in order to survive the above-described stresses, and, in particular, to high ethanol concentration and low pH value [69]. Ethanol in base wines affects yeast growth, viability, and ability to carry on the secondary fermentation [70]. Analysis for ethanol tolerance involves the yeast exposure to several increasing concentrations of ethanol and the monitoring of its growth [71]. It is generally recognized that yeasts should be adapted prior to inoculation for secondary fermentation, by exposing them to increasing trends of ethanol concentrations [72]. This step (known also as *"prise de mousse"*) is essential for an efficient and successful secondary fermentation because it allows the yeast acclimatization to low pH and high amounts of ethanol [73]. Cell acclimatization can improve cell viability, biomass accumulation, and time required for complete secondary fermentation. In addition, improving tolerance to stress by starter yeast may reduce the production of off flavors [21,74,75]. The low pH value is one of the main negative factors that can affect secondary fermentation of a typical base wine. In fact, base wines contain generally high amounts of organic acids (such as tartaric, malic, succinic, and acetic acids) in the undissociated form (at the common pH), that are susceptible to acidify the yeast cytosol, thus leading to sluggish or stuck fermentation [68]. In particular, the parameter to be considered during the selection procedure is the yeast resistance to high concentrations of acetic acid. In fact, acetic acid combined with ethanol can affect yeast fermentative behavior by decreasing cell pH, fermentation rate, and enolase activity [76].

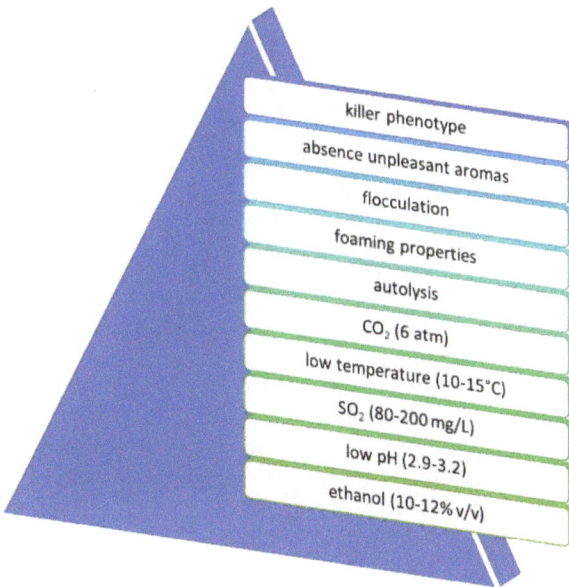

Figure 2. Yeast technological characterization for sparkling wine production.

Another biological parameter that should be considered during the selection of starter cultures for the secondary fermentation is their ability to undergo autolysis. A strain-dependent phenotype that, as a consequence of cell disruption, implies the release in the fermenting wine of several yeast-associated compounds able to influence organoleptic and foaming properties of sparkling wine [77,78]. Several reports suggested the positive effect of yeast strains with high autolytic capacity on sparkling wine quality and foaming properties [24,79,80]. This characteristic is desired at the end of the re-fermentation in sparkling wines, generally two or four months after the end of secondary fermentation. During autolysis, several compounds, such as peptides, amino acids, polysaccharides, higher alcohols, and aldehydes, are released, [69,81,82] and may improve sparkling wine flavor, because some of them (e.g., amino acids) are often precursors of aroma compounds [78]. A further technological characteristic of interest in sparkling wine production is the flocculation capacity of the candidate strain. High flocculation ability is an important criterion among sparkling starter culture selection, because yeast high flocculation aptitude facilitates the removal of sediment at disgorging [23]. At the same time, flocculation prevent yeast sediments from remaining attached to the bottle, thus avoiding turbidity of the final product [7].

Moreover, yeast autolysate compounds could positively influence foaming properties [1] since several macromolecules released by yeast (e.g., mannoproteins) are involved in both foam formation and stabilization [80,83].

Linked to this lytic phenotype, another yeast property to be assessed during starter selection is the presence of the killer phenotype. A killer yeast is a yeast strain which is able to secrete one or more toxic proteins, which are lethal to sensitive yeast strains. This biological phenomenon should be analyzed, checking both killer and sensible phenotypes, because yeast autolysis can be improved with a mixed inoculum of "killer" and sensitive yeast strains for the secondary fermentation [84].

5. Sparkling Wine Production: Role of Non-*Saccharomyces* and Lactic Acid Bacteria

In light of the existing innovative trends in the field of microbial resources in enology, a specific insight is needed for non-*Saccharomyces* yeasts and malolactic lactic acid bacteria (LAB) that might be involved in sparkling wine production [3,85].

Even though non-*Saccharomyces* yeasts are usually involved in winemaking to increase wine's organoleptic properties, little is known on non-*Saccharomyces* application for sparkling wine production. Gonzalez-Royo et al. [9] investigated the chemical and organoleptic properties of base wines accomplished by sequential inoculation of two different non-*Saccharomyces* strains (belonging to *Metschnikowia pulcherrima* and *Torulaspora delbrueckii* species) and *S. cerevisiae* during the alcoholic fermentation of base wines of the AOC Cava. Sequential inoculation of *T. delbrueckii* and *S. cerevisiae* on base wines led to higher glycerol content, lower volatile acidity, and higher foaming properties than their corresponding control wines, suggesting their potential application to innovate specific sparkling wine production [9].

Sequential or co-inoculation of non-*Saccharomyces* and *S. cerevisiae* could be a powerful tool to make base wines with different organoleptic properties, in particular following recent trends on winemaking that suggest the important properties of non-*Saccharomyces* in order to solve specific technological challenges and/or to differentiate the production in terms of sensorial quality [3–6,53].

Among non-*Saccharomyces* yeasts, *Schizosaccharomyces pombe*, usually recognized as a spoilage yeast, presents a malic dehydrogenase activity [86,87] that might have a role in the induction of yeast autolysis during sparkling wine production, enhancing mannoprotein and polysaccharide release during sparkling wine aging [86,87]. Although non-*Saccharomyces* biodiversity could represent a considerable resource of innovation for sparkling wines production, further investigations are needed to understand the possible role of non-*Saccharomyces* starters in the base wine re-fermentation and sparkling wine aging.

The malolactic fermentation (MLF), i.e., the decarboxylation of L-malic acid into L-lactic acid, if desired, takes place post-first fermentation or simultaneously with alcoholic fermentation (co-fermentation of malolactic LAB with yeasts). MLF attenuates acidity, enhances wine biological stability, and modifies sparkling wine sensorial qualities [88]. Several LAB are involved in MLF, the most important is *Oenococcus oeni*, while other LAB species can produce off-flavors and, for this reason, are considered spoilage LAB [89]. The induction of MLF during sparkling wine aging is a common practice in the Champagne region and it allows the production of wine denoted by a higher pH value that also has a reduction in the time requested for their maturation [90]. However, malic and lactic acids were shown to have controversial effects on sparkling wine foaming properties. In fact, malic acid can improve foaming height, while lactic acid produces an opposite action [91–93].

6. Safety Aspects Correlated to Base and Sparkling Wine

Base and sparkling wine safety can be affected by several compounds derived from grapes (e.g., pesticides, phytosanitary products, or trace metal compounds) or from microbial metabolism (e.g., biogenic amines (BA) and the mycotoxin ochratoxin A) [94]. Concerning the compounds of microbial origin, BA are low molecular weight compounds formed in foods by fermentative processes and during aging and storage as a consequence of microbial amino acid decarboxylation [55,95,96]. Sensitive consumers can be intoxicated by BA, which produce several physiological and toxic effects on human health, such as rash, edema, headaches, hypotension, vomiting, palpitations, diarrhea, and heart problems. BA are usually recovered from all fermented foods. Nevertheless, in alcoholic drinks, especially in wine, ethanol and acetaldehyde can enhance the negative effects of biogenic amines by affecting the efficiency of their detoxification by the human body [97–99]. In wine, putrescine represents the major biogenic amine, followed by histamine, tyramine, and cadaverine [100–102], and several authors suggested that BA are produced by LAB metabolism during MLF [98,103–105]. Nevertheless, recent studies also demonstrated that some yeast strains are able to produce BA in wines [53,106,107]. From this point of view, the selection of suitable yeast might represent a fundamental phase to 'build' wine safety.

Another toxic microbial by-product is ocratoxin A (OTA), one of the most common naturally occurring mycotoxins in wine, with several toxic effects [108]. OTA is present in grape musts as a consequence of fungal growth on grapes, particularly attributed to *Aspergillus* and *Penicillium* metabolism [109]. Several OTA-elimination methods have been proposed, from physical to biotechnological mechanisms. The latter represent the best methods to remove OTA from wine without affecting wine organoleptic properties or using toxic chemical compound [110]. Among biologically-based approaches, several studies suggested that *Saccharomyces* and non-*Saccharomyces* strains can remove OTA from wines [111–117].

Therefore, it is very important to monitor the presence of indigenous yeast or LAB during sparkling wine production to reduce the risks to consumer health due to the presence of this toxic microbial compound [118].

7. Biotechnological Applications

Several authors studied biotechnological applications to improve sparkling wine quality, with particular attention to autolysis, flocculation, and sparkling wine flavor [12,14,72,119]. The yeast autolysis, for example, is a slow process that can occur in a few months, or several years, thanks to different environmental conditions, such as temperature, pH, ethanol concentration, nutrient availability, redox potential, and yeast strain [14]. Usually, a slow sparkling wine aging corresponds to an increase of wine organoleptic properties; nevertheless, a long sparkling wine aging can affect entrepreneurial costs. However, several methods, such as adding yeast autolysates to sparkling wine, increasing the aging temperature to accelerate yeast autolysis or microbial genetic improvement, are reported [8,78,84,120]. In particular, the use of yeast autolysates or the increase of aging temperature can affect sparkling wine organoleptic properties [121], while a mixture of killer and killer-sensitive yeast seems to be more promising [32]. In fact, co-inoculation of killer and sensitive *S. cerevisiae* strains allows for increased autolysis, shortening sparkling wine aging time without affecting wine flavor [14,17,32,84]. Genetic improvement can help to design yeast strains with an increased autolytic and flocculation capacity [23]. This method allows improving autolysis and to shorten aging, without affecting wine flavor and needing to introduce modifications in the production process. Among genetic-based methods, the main techniques used are random mutagenesis (for example UV mutagenesis) and genetic engineering [24]. In addition, studies reported in the literature proposed several methods for the development of new *S. cerevisiae* strains with improved flocculation aptitudes, such as clonal selection, recombinant DNA, and hybridization [57,122–124].

With this concern, it is interesting to underline the possibility to hybridize *S. cerevisiae* and *S. uvarum* (*S. bayanus* var. *uvarum*) that led to hybrids with improved technological characteristics (higher fermentative rate, tolerance to low and high temperature, better flocculation capacity, excellent aromatic properties) compared to those of the parental strains [124–128].

8. Conclusions

Both traditional and Charmat methods for sparkling wine productions can benefit from using appropriate starter cultures that could allow the increase of both production efficiency and product quality. Important advances have already been performed by investigating the benefit associated with microbe biodiversity in the vineyard/wine environments. Furthermore, the sector of sparkling wine production will benefit from the development of novel procedures for a renewed exploitation of the enormous opportunities associated with natural microbial biodiversity.

Acknowledgments: This research was supported by the Apulia Region with a grant from Project "Innovazioni di processo e di prodotto nel comparto dei vini spumanti da vitigni autoctoni pugliesi"—IPROVISP (Bando "Aiuti a Sostegno Cluster Tecnologici Regionali"; Project code VJBKVF4). Vittorio Capozzi was supported by 'Fondo di Sviluppo e Coesione 2007–2013—APQ Ricerca Regione Puglia "Programma regionale a sostegno della specializzazione intelligente e della sostenibilità sociale ed ambientale—FutureInResearch".

Conflicts of Interest: The authors declare no conflict of interest.

References

1. Blasco, L.; Viñas, M.; Villa, T.G. Proteins influencing foam formation in wine and beer: The role of yeast. *Int. Microbiol.* **2011**, *14*, 61–71. [PubMed]
2. Legras, J.-L.; Merdinoglu, D.; Cornuet, J.-M.; Karst, F. Bread, beer and wine: *Saccharomyces cerevisiae* diversity reflects human history. *Mol. Ecol.* **2007**, *16*, 2091–2102. [CrossRef] [PubMed]
3. Jolly, N.P.; Varela, C.; Pretorius, I.S. Not your ordinary yeast: Non-*Saccharomyces* yeasts in wine production uncovered. *FEMS Yeast Res.* **2014**, *14*, 215–237. [CrossRef] [PubMed]
4. Garofalo, C.; El Khoury, M.; Lucas, P.; Bely, M.; Russo, P.; Spano, G.; Capozzi, V. Autochthonous starter cultures and indigenous grape variety for regional wine production. *J. Appl. Microbiol.* **2015**, *118*, 1395–1408. [CrossRef] [PubMed]
5. Garofalo, C.; Tristezza, M.; Grieco, F.; Spano, G.; Capozzi, V. From grape berries to wine: Population dynamics of cultivable yeasts associated to "Nero di Troia" autochthonous grape cultivar. *World J. Microbiol. Biotechnol.* **2016**, *32*, 59. [CrossRef] [PubMed]
6. Capozzi, V.; Garofalo, C.; Chiriatti, M.A.; Grieco, F.; Spano, G. Microbial terroir and food innovation: The case of yeast biodiversity in wine. *Microbiol. Res.* **2015**, *181*, 75–83. [CrossRef] [PubMed]
7. Carrascosa, A.V.; Martinez-Rodiguez, A.; Cebollero, E.; Gonzalez, R. *Saccharomyces* Yeasts II: Secondary fermentation. In *Molecular Wine Microbiology*; Elsevier: London, UK, 2011; Volume 2, pp. 33–48.
8. Cebollero, E.; Gonzalez, R. Induction of autophagy by second-fermentation yeasts during elaboration of sparkling wines. *Appl. Environ. Microbiol.* **2006**, *72*, 4121–4127. [CrossRef] [PubMed]
9. González-Royo, E.; Pascual, O.; Kontoudakis, N.; Esteruelas, M.; Esteve-Zarzoso, B.; Mas, A.; Canals, J.M.; Zamora, F. Oenological consequences of sequential inoculation with non-*Saccharomyces* yeasts (*Torulaspora delbrueckii* or *Metschnikowia pulcherrima*) and *Saccharomyces cerevisiae* in base wine for sparkling wine production. *Eur. Food Res. Technol.* **2014**, *240*, 999–1012. [CrossRef]
10. Ody-Brasier, A.; Vermeulen, F. The price you pay price-setting as a response to norm violations in the market for champagne grapes. *Adm. Sci. Q.* **2014**, *59*, 109–144. [CrossRef]
11. Torresi, S.; Frangipane, M.T.; Anelli, G. Biotechnologies in sparkling wine production. Interesting approaches for quality improvement: A review. *Food Chem.* **2011**, *129*, 1232–1241. [CrossRef] [PubMed]
12. Perpetuini, G.; Di Gianvito, P.; Arfelli, G.; Schirone, M.; Corsetti, A.; Tofalo, R.; Suzzi, G. Biodiversity of autolytic ability in flocculent *Saccharomyces cerevisiae* strains suitable for traditional sparkling wine fermentation. *Yeast* **2016**, *33*, 303–312. [CrossRef] [PubMed]
13. Pérez-Magariño, S.; Martínez-Lapuente, L.; Bueno-Herrera, M.; Ortega-Heras, M.; Guadalupe, Z.; Ayestarán, B. Use of commercial dry yeast products rich in mannoproteins for white and rosé sparkling wine elaboration. *J. Agric. Food Chem.* **2015**, *63*, 5670–5681. [CrossRef] [PubMed]
14. Lombardi, S.J.; de Leonardis, A.; Lustrato, G.; Testa, B.; Iorizzo, M. Yeast autolysis in sparkling wine aging: Use of killer and sensitive *Saccharomyces cerevisiae* strains in co-culture. *Recent Pat. Biotechnol.* **2015**, *9*, 223–230. [CrossRef] [PubMed]

15. Alexandre, H.; Guilloux-Benatier, M. Yeast autolysis in sparkling wine—A review. *Aust. J. Grape Wine Res.* **2006**, *12*, 119–127. [CrossRef]
16. Tita, O.; Jascanu, V.; Tita, M.; Sand, C. Economical comparative analysis of different bottle fermentation methods. In Proceedings of the AVA 2003, International Conference on Agricultural Economics, Rural Development and Informatics in the New Millennium, Debrecen, Hungary, 1–2 April 2003.
17. Pozo-Bayón, M.A.; Andujar-Ortiz, I.; Alcaide-Hidalgo, J.M.; Martín-Alvarez, P.J.; Moreno-Arribas, M.V. Characterization of commercial inactive dry yeast preparations for enological use based on their ability to release soluble compounds and their behavior toward aroma compounds in model wines. *J. Agric. Food Chem.* **2009**, *57*, 10784–10792. [CrossRef] [PubMed]
18. Ganss, S.; Kirsch, F.; Winterhalter, P.; Fischer, U.; Schmarr, H.-G. Aroma changes due to second fermentation and glycosylated precursors in Chardonnay and Riesling sparkling wines. *J. Agric. Food Chem.* **2011**, *59*, 2524–2533. [CrossRef] [PubMed]
19. Pozo-Bayón, M.A.; Hernández, M.T.; Martín-Alvarez, P.J.; Polo, M.C. Study of low molecular weight phenolic compounds during the aging of sparkling wines manufactured with red and white grape varieties. *J. Agric. Food Chem.* **2003**, *51*, 2089–2095. [CrossRef] [PubMed]
20. Ribéreau-Gayon, P.; Dubourdieu, D.; Donèche, B.; Lonvaud, A. *Trattato di Enologia I, Microbiologia del Vino*; Edagricole: Bologna, Italy, 2004.
21. Zoecklein, B. *A Review of Méthode Champenoise Production*, 2nd ed.; Virginia Tech: Blacksburg, VA, USA, 2002.
22. Martínez-Rodríguez, A.J.; Pueyo, E. Sparkling Wines and Yeast Autolysis. In *Wine Chemistry and Biochemistry*; Moreno-Arribas, M.V., Polo, M.C., Eds.; Springer: New York, NY, USA, 2009.
23. Bidan, P.; Feuillat, M.; Moulin, J.P. Les vins mousseux. *Bull. l'OIV* **1986**, *59*, 663–664.
24. Nunez, Y.P.; Carrascosa, A.V.; González, R.; Polo, M.C.; Martínez-Rodríguez, A.J. Effect of accelerated autolysis of yeast on the composition and foaming properties of sparkling wines elaborated by a champenoise method. *J. Agric. Food Chem.* **2005**, *53*, 7232–7237. [CrossRef] [PubMed]
25. Luguera, C.; Moreno-Arribas, M.V.; Pueyo, E.; Bartolomé, B.; Polo, M.C. Fractionation and partial characterization of protein fractions present at different stages of the production of sparkling wines. *Food Chem.* **1998**, *63*, 465–471. [CrossRef]
26. Nikolaou, E.; Soufleros, E.H.; Bouloumpasi, E.; Tzanetakis, N. Selection of indigenous *Saccharomyces cerevisiae* strains according to their oenological characteristics and vinification results. *Food Microbiol.* **2006**, *23*, 205–211. [CrossRef] [PubMed]
27. Šuranská, H.; Vránová, D.; Omelková, J. Isolation, identification and characterization of regional indigenous *Saccharomyces cerevisiae* strains. *Braz. J. Microbiol.* **2016**, *47*, 181–190. [CrossRef] [PubMed]
28. Mercado, L.; Sturm, M.E.; Rojo, M.C.; Ciklic, I.; Martínez, C.; Combina, M. Biodiversity of *Saccharomyces cerevisiae* populations in Malbec vineyards from the "Zona Alta del Río Mendoza" region in Argentina. *Int. J. Food Microbiol.* **2011**, *151*, 319–326. [CrossRef] [PubMed]
29. Pérez, F.; Regodón, J.; Valdés, M.; de Miguel, C.; Ramírez, M. Cycloheximide resistance as marker for monitoring yeasts in wine fermentations. *Food Microbiol.* **2000**, *17*, 119–128. [CrossRef]
30. Ambrona, J.; Maqueda, M.; Zamora, E.; Ramírez, M. Sulfometuron Methyl Resistance as Genetic Marker for Monitoring Yeast Populations in Wine Fermentations. *J. Agric. Food Chem.* **2005**, *53*, 7438–7443. [CrossRef] [PubMed]
31. Ambrona, J.; Vinagre, A.; Maqueda, M.; Ramírez, M. Rhodamine-pink as a genetic marker for yeast populations in wine fermentation. *J. Agric. Food Chem.* **2006**, *54*, 2977–2984. [CrossRef] [PubMed]
32. Velázquez, R.; Zamora, E.; Álvarez, M.L.; Álvarez, M.L.; Ramírez, M. Using mixed inocula of new killer strains of *Saccharomyces cerevisiae* to improve the quality of traditional sparkling-wine. *Food Microbiol.* **2016**, *59*, 150–160. [CrossRef] [PubMed]
33. Martínez, C.; Cosgaya, P.; Vásquez, C.; Gac, S.; Ganga, A. High degree of correlation between 464 molecular polymorphism and geographic origin of wine yeast strains. *J. Appl. Microbiol.* **2007**, *103*, 2185–2195. [CrossRef] [PubMed]
34. Schuller, D.; Valero, E.; Dequin, S.; Casal, M. Survey of molecular methods for the typing of wine yeast strains. *FEMS Microbiol. Lett.* **2004**, *231*, 19–26. [CrossRef]
35. Querol, A.; Barrio, E.; Huerta, T.; Ramón, D. Molecular monitoring of wine fermentations conducted by active dry yeast strains. *Appl. Environ. Microbiol.* **1992**, *58*, 2948–2953. [PubMed]

36. Guillamon, J.M.; Sabaté, J.; Barrio, E.; Cano, J.; Querol, A. Rapid identification of wine yeast species based on RFLP analysis of the ribosomal internal transcribed spacer (ITS) region. *Arch. Microbiol.* **1998**, *169*, 387–392. [CrossRef] [PubMed]

37. Fernández-Espinar, M.T.; López, V.; Ramón, D.; Bartra, E.; Querol, A. Study of the authenticity of commercial wine yeast strains by molecular techniques. *Int. J. Food Microbiol.* **2001**, *70*, 1–10. [CrossRef]

38. Torija, M.J.; Rozès, N.; Poblet, M.; Guillamón, J.M.; Mas, A. Yeast population dynamics in spontaneous fermentations: Comparison between two different wine-producing areas over a period of three years. *Antonie Van Leeuwenhoek* **2001**, *79*, 345–352. [CrossRef] [PubMed]

39. Blondin, B.; Vezinhet, F. Identification de souches de levures oenologiques par leurs caryotypes obtenus en électrophorèse en champ pulsé. *Rev. Franç. Oenol.* **1988**, *28*, 7–11.

40. Ness, F.; Lavallée, F.; Dubourdieu, D.; Aigle, M.; Dulau, L. Identification of yeast strains using the polymerase chain reaction. *J. Sci. Food Agric.* **1993**, *62*, 89–94. [CrossRef]

41. Quesada, M.P.; Cenis, J.L. Use of Random Amplified Polymorphic DNA (RAPD-PCR) in the Characterization of Wine Yeasts. *Am. J. Enol. Vitic.* **1995**, *46*, 204–208.

42. Strand, M.; Prolla, T.A.; Liskay, R.M.; Petes, T.D. Destabilization of tracts of simple repetitive DNA in yeast by mutations affecting DNA mismatch repair. *Nature* **1993**, *365*, 274–276. [CrossRef] [PubMed]

43. Pérez, M.A.; Gallego, F.J.; Martínez, I.; Hidalgo, P. Detection, distribution and selection of microsatellites (SSRs) in the genome of the yeast *Saccharomyces cerevisiae* as molecular markers. *Lett. Appl. Microbiol.* **2001**, *33*, 461–466. [CrossRef] [PubMed]

44. Fay, J.C.; Benavides, J.A. Evidence for domesticated and wild populations of *Saccharomyces cerevisiae*. *PLoS Genet.* **2005**, *1*, 66–71. [CrossRef] [PubMed]

45. Guillamón, J.M.; Barrio, E.; Querol, A. Characterization of Wine Yeast Strains of the *Saccharomyces* genus on the basis of molecular markers: Relationships between genetic distance and geographic or ecological origin. *Syst. Appl. Microbiol.* **1996**, *19*, 122–132. [CrossRef]

46. Capece, A.; Romaniello, R.; Siesto, G.; Pietrafesa, R.; Massari, C.; Poeta, C.; Romano, P. Selection of indigenous *Saccharomyces cerevisiae* strains for Nero d'Avola wine and evaluation of selected starter implantation in pilot fermentation. *Int. J. Food Microbiol.* **2010**, *144*, 187–192. [CrossRef] [PubMed]

47. Capece, A.; Romaniello, R.; Siesto, G.; Romano, P. Diversity of *Saccharomyces cerevisiae* yeasts associated to spontaneously fermenting grapes from an Italian "heroic vine-growing area". *Food Microbiol.* **2012**, *31*, 159–166. [CrossRef] [PubMed]

48. Barrajón, N.; Arévalo-Villena, M.; Rodríguez-Aragón, L.J.; Briones, A. Ecological study of wine yeast in inoculated vats from La Mancha region. *Food Control* **2009**, *20*, 778–783. [CrossRef]

49. Maqueda, M.; Zamora, E.; Rodríguez-Cousiño, N.; Ramírez, M. Wine yeast molecular typing using a simplified method for simultaneously extracting mtDNA, nuclear DNA and virus dsRNA. *Food Microbiol.* **2010**, *27*, 205–209. [CrossRef] [PubMed]

50. Muñoz, R.; Gómez, A.; Robles, V.; Rodríguez, P.; Cebollero, E.; Tabera, L.; Carrascosa, A.V.; Gonzalez, R. Multilocus sequence typing of oenological *Saccharomyces cerevisiae* strains. *Food Microbiol.* **2009**, *26*, 841–846. [CrossRef] [PubMed]

51. Tristezza, M.; Fantastico, L.; Vetrano, C.; Bleve, G.; Corallo, D.; Grieco, F.; Mita, G.; Grieco, F. Molecular and technological characterization of *Saccharomyces cerevisiae* strains isolated from natural fermentation of Susumaniello grape must in Apulia, Southern Italy. *Int. J. Microbiol.* **2014**, *2014*, 897427. [CrossRef] [PubMed]

52. Tristezza, M.; Vetrano, C.; Bleve, G.; Grieco, F.; Tufariello, M.; Quarta, A.; Mita, G.; Spano, G.; Grieco, F. Autochthonous fermentation starters for the industrial production of Negroamaro wines. *J. Ind. Microbiol. Biotechnol.* **2011**, *39*, 81–92. [CrossRef] [PubMed]

53. Tristezza, M.; Vetrano, C.; Bleve, G.; Spano, G.; Capozzi, V.; Logrieco, A.; Mita, G.; Grieco, F. Biodiversity and safety aspects of yeast strains characterized from vineyards and spontaneous fermentations in the Apulia Region, Italy. *Food Microbiol.* **2013**, *36*, 335–342. [CrossRef] [PubMed]

54. Pretorius, I.S. Tailoring wine yeast for the new millennium: Novel approaches to the ancient art of winemaking. *Yeast* **2000**, *16*, 675–729. [CrossRef]

55. Cappello, M.S.; Poltronieri, P.; Blaiotta, G.; Zacheo, G. Molecular and physiological characteristics of a grape yeast strain containing atypical genetic material. *Int. J. Food Microbiol.* **2010**, *144*, 72–80. [CrossRef] [PubMed]

56. Bely, M.; Stoeckle, P.; Masneuf-Pomarède, I.; Dubourdieu, D. Impact of mixed *Torulaspora delbrueckii-Saccharomyces cerevisiae* culture on high-sugar fermentation. *Int. J. Food Microbiol.* **2008**, *122*, 312–320. [CrossRef] [PubMed]
57. Rainieri, S.; Pretorius, I.S. Selection and improvement of wine yeasts. *Ann. Microbiol.* **2000**, *50*, 15–32.
58. Henschke, P.A. Wine yeast. In *Yeast Sugar Metabolism*; Zimmermann, F.K., Entian, K.D., Eds.; Technomic Publishing Co Inc.: Lancaster, PA, USA, 1997; pp. 527–560.
59. Salmon, J.M.; Vincent, O.; Mauricio, J.C.; Bely, M.; Barre, P. Sugar transport inhibition and apparent loss of activity in *Saccharomyces cerevisiae* as a major limiting factor of enological fermentations. *Am. J. Enol. Vitic.* **1993**, *44*, 56–64.
60. Blomberg, A.; Adler, L. Physiology of osmotolerance in fungi. *Adv. Microb. Physiol.* **1992**, *33*, 145–212. [PubMed]
61. Romano, P. Lievito starter e qualità aromatica del vino. *Inf. Agric.* **2006**, *62*, 27–31.
62. Csoma, H.; Zakany, N.; Capece, A.; Romano, P.; Sipiczki, M. Biological diversity of *Saccharomyces* yeasts of spontaneously fermenting wines in four wine regions: Comparative genotypic and phenotypic analysis. *Int. J. Food Microbiol.* **2010**, *140*, 239–248. [CrossRef] [PubMed]
63. Romano, P.; Fiore, C.; Paraggio, M.; Caruso, M.; Capece, A. Function of yeast species and strains in wine flavour. *Int. J. Food Microbiol.* **2003**, *86*, 169–180. [CrossRef]
64. Manzanares, P.; Ramón, D.; Querol, A. Screening of non-*Saccharomyces* wine yeasts for the production of b-D-xylosidase activity. *Int. J. Food Microbiol.* **1999**, *46*, 105–112. [CrossRef]
65. Beltran, G.; Torija, M.J.; Novo, M.; Ferrer, N.; Poblet, M.; Guillamón, J.M.; Rozès, N.; Mas, A. Analysis of yeast populations during alcoholic fermentation: A six year follow-up study. *Syst. Appl. Microbiol.* **2002**, *25*, 287–293. [CrossRef] [PubMed]
66. Cocolin, L.; Bisson, L.F.; Mills, D.A. Direct profiling of the yeast dynamics in wine fermentations. *FEMS Microbiol. Lett.* **2000**, *189*, 81–87. [CrossRef] [PubMed]
67. Santamaría, P.; Garijo, P.; López, R.; Tenorio, C.; Rosa Gutiérrez, A. Analysis of yeast population during spontaneous alcoholic fermentation: Effect of the age of the cellar and the practice of inoculation. *Int. J. Food Microbiol.* **2005**, *103*, 49–56. [CrossRef] [PubMed]
68. Borrull, A.; Poblet, M.; Rozès, N. New insights into the capacity of commercial wine yeasts to grow on sparkling wine media. Factor screening for improving wine yeast selection. *Food Microbiol.* **2015**, *48*, 41–48. [CrossRef] [PubMed]
69. Kemp, B.; Alexandre, H.; Robillard, B.; Marchal, R. Effect of production phase on bottle-fermented sparkling wine quality. *J. Agric. Food Chem.* **2015**, *63*, 19–38. [CrossRef] [PubMed]
70. Kumar, G.R.; Goyashiki, R.; Ramakrishnan, V.; Karpel, J.E.; Bisson, L.F. Genes required for ethanol tolerance and utilization in *Saccharomyces cerevisiae*. *Am. J. Enol. Vitic.* **2008**, *59*, 401–411.
71. Bisson, L.F.; Block, D.E. Ethanol tolerance in *Saccharomyces*. In *Biodiversity and Biotechnology of Wine Yeasts*; Ciani, M., Ed.; Research Signpost: Kerala, India, 2002.
72. Martí-Raga, M.; Martín, V.; Gil, M.; Sancho, M.; Zamora, F.; Mas, A.; Beltran, G. Contribution of yeast and base wine supplementation to sparkling wine composition. *J. Sci. Food Agric.* **2016**. [CrossRef] [PubMed]
73. Laurent, M.; Valade, M. La propagation des levains de tirage. *Vign. Champen.* **1998**, *3*, 29–52.
74. Monk, P.R.; Storer, R.J. The kinetics of yeast growth and sugar utilization in tirage: The influence of different methods of starter culture preparation and inoculation levels. *Am. J. Enol. Vitic.* **1986**, *37*, 72–76.
75. Benucci, I.; Liburdi, K.; Cerreti, M.; Esti, M. Characterization of active dry wine yeast during starter culture (Pied de cuve). Preparation for sparkling wine production. *J. Food Sci.* **2016**, *81*, 2015–2020. [CrossRef] [PubMed]
76. Pampulha, M.E.; Loureiro-Dias, M.C. Activity of glycolytic enzymes of *Saccharomyces cerevisiae* in the presence of acetic acid. *Appl. Microbiol. Biotechnol.* **1989**, *34*, 375–380. [CrossRef]
77. Martínez-Rodríguez, A.J.; Polo, M.C. Characterization of the Nitrogen Compounds Released during Yeast Autolysis in a Model Wine System. *J. Agric. Food Chem.* **2000**, *48*, 1081–1085. [CrossRef] [PubMed]
78. Charpentier, C.; Feuillat, M. Yeast autolysis. In *Wine Microbiology and Biotechnology*; Fleet, G.H., Ed.; Harwood Academic Publishers: Chur, Switzerland, 1993; pp. 225–242.
79. Martínez-Rodríguez, A.J.; Polo, M.C.; Carrascosa, A.V. Structural and ultrastructural changes in yeast cells during autolysis in a model wine system and in sparkling wines. *Int. J. Food Microbiol.* **2001**, *71*, 45–51. [CrossRef]

80. Martínez-Rodriguez, A.J.; Carrascosa, A.V.; Polo, M.C. Release of nitrogen compounds to the extracellular medium by three strains of *Saccharomyces cerevisiae* during induced autolysis in a model wine system. *Int. J. Food Microbiol.* **2001**, *68*, 155–160. [CrossRef]

81. Molnar, I.; Oura, E.; Suomalainen, H. Study of volatile substrates produced during the autolysis of champagne yeast. *Acta Aliment.* **1981**, *10*, 27–36.

82. Torresi, S.; Frangipane, M.T.; Garzillo, A.M.V.; Massantini, R.; Contini, M. Effects of a β-glucanase enzymatic preparation on yeast lysis during aging of traditional sparkling wines. *Food Res. Int.* **2014**, *55*, 83–92. [CrossRef]

83. Moreno-Arribas, M.V.; Polo, M.C. Amino Acids and Biogenic Amines. In *Wine Chemistry and Biochemistry*; Moreno-Arribas, M.V., Polo, M.C., Eds.; Springer: New York, NY, USA, 2009; pp. 163–189.

84. Todd, B.E.N.; Fleet, G.H.; Henschke, P.A. Promotion of autolysis through the interaction of killer and sensitive yeasts: Potential application in sparkling wine production. *Am. J. Enol. Vitic.* **2000**, *51*, 65–72.

85. Jolly, J.; Augustyn, O.P.H.; Pretorius, I.S. The role and use of non-Saccharomyces yeasts in wine production. *S. Afr. J. Enol. Vitic.* **2006**, *27*, 15–39.

86. Suárez-Lepe, J.A.; Morata, A. New trends in yeast selection for winemaking. *Trends Food Sci. Technol.* **2012**, *23*, 39–50. [CrossRef]

87. Benito, S.; Gálvez, L.; Palomero, F.; Calderón, F.; Morata, A.; Palmero, D.; Suárez-Lepe, J.A. *Schizosaccharomyces* selective differential media. *Afr. J. Microbiol. Res.* **2013**, *7*, 3026–3036.

88. Lasik, M. The application of malolactic fermentation process to create good-quality grape wine produced in cool-climate countries: A review. *Eur. Food Res. Technol.* **2013**, *237*, 843–850. [CrossRef]

89. Bartowsky, E.J.; Borneman, A.R. Genomic variations of *Oenococcus oeni* strains and the potential to impact on malolactic fermentation and aroma compounds in wine. *Appl. Microbiol. Biotechnol.* **2011**, *92*, 441–447. [CrossRef] [PubMed]

90. Auge, D.; Valade, M.; Moncomble, D. Acidity of Champagne wines: Use of malolactic fermentation? *Vign. Champen. Epernay* **2000**, *121*, 44–56.

91. Girbau-Solà, T.; López-Barajas, M.; López-Tamames, E.; Buxaderas, S. Foam aptitude of Trepat and Monastrell red varieties in cava elaboration. 2. Second fermentation and aging. *J. Agric. Food Chem.* **2002**, *50*, 5600–5604. [CrossRef] [PubMed]

92. Girbau-Solà, T.; López-Tamames, E.; Buján, J.; Buxaderas, S. Foam aptitude of Trepat and Monastrell red varieties in cava elaboration. 1. Base wine characteristics. *J. Agric. Food Chem.* **2002**, *50*, 5596–5599. [CrossRef] [PubMed]

93. Lopez-Barajas, M.; Lopez-Tamames, E.; Buxaderas, S.; de la Torre-Boronat, M.C. Effect of vinification and variety on foam capacity of wine. *Am. J. Enol. Vitic.* **1998**, *49*, 397–402.

94. Pozo-Bayón, M.Á.; Monagas, M.; Bartolomé, B.; Moreno-Arribas, M.V. Wine features related to safety and consumer health: An integrated perspective. *Crit. Rev. Food Sci. Nutr.* **2012**, *52*, 31–54. [CrossRef] [PubMed]

95. Ferreira, I.M.; Pinho, O. Biogenic amines in Portuguese traditional foods and wines. *J. Food Prot.* **2006**, *69*, 2293–2303. [CrossRef] [PubMed]

96. Ancín-Azpilicueta, C.; González-Marco, A.; Jiménez-Moreno, N. Current knowledge about the presence of amines in wine. *Crit. Rev. Food Sci. Nutr.* **2008**, *48*, 257–275. [CrossRef] [PubMed]

97. Maynard, L.S.; Schenker, V.J. Monoamine-oxidase inhibition by ethanol in vitro. *Nature* **1996**, *196*, 575. [CrossRef]

98. Coton, M.; Romano, A.; Spano, G.; Ziegler, K.; Vetrana, C.; Desmarais, C.; Lonvaud-Funel, A.; Lucas, P.; Coton, E. Occurrence of biogenic amine-forming lactic acid bacteria in wine and cider. *Food Microbiol.* **2010**, *27*, 1078–1085. [CrossRef] [PubMed]

99. Landete, J.M.; de Las Rivas, B.; Marcobal, A.; Muñoz, R. Molecular methods for the detection of biogenic amine-producing bacteria on foods. *Int. J. Food Microbiol.* **2007**, *117*, 258–269. [CrossRef] [PubMed]

100. Spano, G.; Russo, P.; Lonvaud-Funel, A.; Lucas, P.; Alexandre, H.; Grandvalet, C.; Coton, E.; Coton, M.; Barnavon, L.; Bach, B.; et al. Biogenic amines in fermented foods. *Eur. J. Clin. Nutr.* **2010**, *64*, S95–S100. [CrossRef] [PubMed]

101. Beneduce, L.; Romano, A.; Capozzi, V.; Lucas, P.; Barnavon, L.; Bach, B.; Vuchot, P.; Grieco, F.; Spano, G. Biogenic amine in wines. *Ann. Microbiol.* **2010**, *60*, 573–578. [CrossRef]

102. Landete, J.M.; Ferrer, S.; Polo, L.; Pardo, I. Biogenic Amines in Wines from Three Spanish Regions. *J. Agric. Food Chem.* **2005**, *53*, 1119–1124. [CrossRef] [PubMed]

103. Smit, A.; Moses, S.G.; Pretorius, I.S.; Cordero Otero, R.R. The Thr505 and Ser557 residues of the AGT1-encoded alpha-glucoside transporter are critical for maltotriose transport in *Saccharomyces cerevisiae*. *J. Appl. Microbiol.* **2008**, *104*, 1103–1111. [CrossRef] [PubMed]

104. Marcobal, A.; Martín-Alvarez, P.J.; Polo, M.C.; Muñoz, R.; Moreno-Arribas, M.V. Formation of biogenic amines throughout the industrial manufacture of red wine. *J. Food Prot.* **2006**, *69*, 397–404. [CrossRef] [PubMed]

105. Torrea, D.; Ancín, C. Content of biogenic amines in a Chardonnay wine obtained through spontaneous and inoculated fermentations. *J. Agric. Food Chem.* **2002**, *50*, 4895–4899. [CrossRef] [PubMed]

106. Caruso, M.; Capece, A.; Salzano, G.; Romano, P. Typing of *Saccharomyces cerevisiae* and *Kloeckera apiculata* strains from Aglianico wine. *Lett. Appl. Microbiol.* **2002**, *34*, 323–328. [CrossRef] [PubMed]

107. Granchi, L.; Romano, P.; Mangani, S.; Guerrini, S.; Vincenzini, M. Production of biogenic amines by wine microorganisms. *Bull. l'OIV Off. Int. Vigne Vin* **2005**, *78*, 595–610.

108. Delage, N.; d'Harlingue, A.; Colonna Ceccaldi, B.; Bompeix, G. Occurrence of mycotoxins in fruit juices and wine. *Food Control* **2003**, *14*, 225–227. [CrossRef]

109. Esti, M.; Benucci, I.; Liburdi, K.; Acciaro, G. Monitoring of ochratoxin A fate during alcoholic fermentation of wine-must. *Food Control* **2012**, *27*, 53–56. [CrossRef]

110. Anli, E.; Bayram, M. Ochratoxin A in Wines. *Food Rev. Int.* **2009**, *25*, 214–232. [CrossRef]

111. Angioni, A.; Caboni, P.; Garau, A.; Farris, A.; Orro, D.; Budroni, M.; Cabras, P. In vitro interaction between ochratoxin A and different strains of *Saccharomyces cerevisiae* and *Kloeckera apiculata*. *J. Agric. Food Chem.* **2007**, *55*, 2043–2048. [CrossRef] [PubMed]

112. Caridi, A. Enological functions of parietal yeast mannoproteins. *Antonie Van Leeuwenhoek* **2006**, *89*, 417–422. [CrossRef] [PubMed]

113. Caridi, A.; Cufari, J.A.; Ramondino, D. Isolation and clonal pre-selection of enological *Saccharomyces*. *J. Gen. Appl. Microbiol.* **2002**, *48*, 261–267. [CrossRef] [PubMed]

114. Csutorás, C.; Rácz, L.; Rácz, K.; Fűtő, P.; Forgó, P.; Kiss, A. Monitoring of ochratoxin A during the fermentation of different wines by applying high toxin concentrations. *Microchem. J.* **2013**, *107*, 182–184. [CrossRef]

115. Meca, G.; Blaiotta, G.; Ritieni, A. Reduction of ochratoxin A during the fermentation of Italian red wine Moscato. *Food Control* **2010**, *21*, 579–583. [CrossRef]

116. Piotrowska, M.; Nowak, A.; Czyzowska, A. Removal of ochratoxin A by wine *Saccharomyces cerevisiae* strains. *Eur. Food Res. Technol.* **2013**, *236*, 441–447. [CrossRef]

117. Petruzzi, L.; Bevilacqua, A.; Corbo, M.R.; Garofalo, C.; Baiano, A.; Sinigaglia, M. Selection of autochthonous *Saccharomyces cerevisiae* strains as wine starters using a polyphasic approach and ochratoxin A Removal. *J. Food Prot.* **2014**, *77*, 1168–1177. [CrossRef] [PubMed]

118. Russo, P.; Capozzi, V.; Spano, G.; Corbo, M.R.; Sinigaglia, M.; Bevilacqua, A. Metabolites of microbial origin with an impact on health: Ochratoxin A and biogenic amines. *Front. Microbiol.* **2016**, *7*, 482. [CrossRef] [PubMed]

119. Penacho, V.; Valero, E.; Gonzalez, R. Transcription profiling of sparkling wine second fermentation. *Int. J. Food Microbiol.* **2012**, *153*, 176–182. [CrossRef] [PubMed]

120. Tini, V.; Zambonelli, C.; Benevelli, M.; Castellari, L. The autolysogenic *Saccharomyces cerevisiae* strains for the sparkling wines production. *Ind. Bevande* **1995**, *24*, 113–118.

121. Peppler, H.J. Yeast extracts. In *Fermented Foods. Economic Microbiology*; Rose, A.H., Ed.; Academic Press: London, UK, 1982; Volume 7, pp. 293–311.

122. Romano, P.; Soli, M.G.; Suzzi, G.; Grazia, L.; Zambonelli, C. Improvement of a Wine *Saccharomyces cerevisiae* Strain by a Breeding Program. *Appl. Environ. Microbiol.* **1985**, *50*, 1064–1067. [PubMed]

123. Giudici, P.; Solieri, L.; Pulvirenti, A.M.; Cassanelli, S. Strategies and perspectives for genetic improvement of wine yeasts. *Appl. Microbiol. Biotechnol.* **2005**, *66*, 622–628. [CrossRef] [PubMed]

124. Coloretti, F.; Zambonelli, C.; Tini, V. Characterization of flocculent *Saccharomyces* interspecific hybrids for the production of sparkling wines. *Food Microbiol.* **2006**, *23*, 672–676. [CrossRef] [PubMed]

125. Winge, O.; Lausten, O. Artificial species hybridisation in yeast. *Comp. Rend. Trav. Lab. Carlsberg. Sér. Physiol.* **1938**, *22*, 235–244.

126. Castellari, M.; Ferruzzi, A.; Magrini, P.; Giudici, P.; Passarelli, C. Zambonelli Unbalanced wine fermentation by cryotolerant vs. non-cryotolerant. *Saccharomyces* strains. *Vitis* **1994**, *33*, 49–52.

127. Massoutier, C.; Alexandre, H.; Feuillat, M.; Charpentier, C. Isolation and characterization of cryotolerant *Saccharomyces* strains. *Vitis* **1998**, *37*, 55–59.

128. Suzzi, G.; Romano, P.; Zambonelli, C. Flocculation of wine yeasts: Frequency, differences, and stability of the character. *Can. J. Microbiol.* **1984**, *30*, 36–39. [CrossRef]

fermentation

MDPI

Article

Purification and Properties of Yeast Proteases Secreted by *Wickerhamomyces anomalus* 227 and *Metschnikovia pulcherrima* 446 during Growth in a White Grape Juice

Martina Schlander [1], Ute Distler [2], Stefan Tenzer [2], Eckhard Thines [1] and Harald Claus [1,*]

[1] Institute for Microbiology and Wine Research, Johannes Gutenberg University of Mainz, Becherweg 15, D-55099 Mainz, Germany; schlande@uni-mainz.de (M.S.); thines@uni-mainz.de (E.T.)

[2] Institute of Immunology, University Medical Centre of the Johannes Gutenberg University Mainz, Langenbeckstr. 1, D-55131 Mainz, Germany; ute.distler@uni-mainz.de (U.D.); tenzer@uni-mainz.de (S.T.)

* Correspondence: hclaus@uni-mainz.de; Tel.: +49-6131-3923542

Academic Editor: Ronnie G. Willaert
Received: 25 October 2016; Accepted: 19 December 2016; Published: 26 December 2016

Abstract: Aspartic proteases are of significant importance for medicine and biotechnology. In spite of sufficient evidence that many non-*Saccharomyces* yeasts produce extracellular proteases, previous research has focused on the enzymes of *Candida* species because of their role as virulence factors. Nowadays, there is also increasing interest for their applications in industrial processes, mainly because of their activities at low pH values. Here, we report the features of new acid proteases isolated from wine-relevant yeasts *Metschnikovia pulcherrima* and *Wickerhamomyces anomalus*. To our knowledge, this is the first detailed description of such an enzyme derived from strains of *W. anomalus*. Deviating to most former studies, we could demonstrate that the yeasts produce these enzymes in a natural substrate (grape juice) during the active growth phase. The enzymes were purified from concentrated grape juice by preparative isoelectric focusing. Biochemical data (maximum activity at ≈ pH 3.0, inhibition by pepstatin A) classify them as aspartic proteases. For *W. anomalus* 227, this assumption was confirmed by the protein sequence of WaAPR1 determined by LC-MS/MS. The sequence revealed a signal peptide for secretion, as well as a peptidase A1 domain with two aspartate residues in the active site. The enzyme has a calculated molecular mass of 47 kDa and an isolelectric point of 4.11.

Keywords: aspartic protease; *Wickerhamomyces*; *Metschnikovia*; grape juice; wine protein

1. Introduction

The use of mixed starters of selected non-*Saccharomyces* yeasts and *Saccharomyces cerevisiae* is of increasing interest for production of novel wines with more complex organoleptic characteristics [1] and/or lower ethanol contents [2]. Non-*Saccharomyces* wine yeasts, also called "wild" yeasts, can enhance the analytical composition and aroma profile of wine by production of secondary metabolites and secretion of enzymes [3–5]. Yeasts of the genera *Kloeckera*, *Candida*, *Debaryomyces*, *Rhodotorula*, *Pichia*, *Wickerhamomyces*, *Zygosaccharomyces*, *Hanseniaspora*, *Kluyveromyces* and *Metschnikowia* produce hydrolytic exoenzymes (esterases, lipases, glycosidases, glucanases, pectinases, amylases, and proteases) that interact with grape compounds [6]. Particularly, glycoside hydrolases can release aroma active compounds in grape must from their odourless glycosidic precursors [7]. Others produce pectinolytic enzymes that can promote grape must clarification and may substitute for fungal enzymes, which are currently used for winemaking [8].

Recent research in wine biotechnology has focused on acid proteases to prevent formation of wine haze. A thermotolerant fungal protease (aspergilloglutamic peptidase) has already been approved for winemaking in Australia [9].

Degradation of haze forming-proteins by enzymes is an attractive alternative to bentonite fining because it would minimize losses of wine volume and aroma. Appropriate proteases must be active under harsh winemaking conditions, i.e., low pH (~3.5), low temperature (~15 °C), presence of ethanol ($\geq$ 10% v/v), phenolic compounds and sulphites. Another problem is the intrinsic stability of haze forming proteins due to their high numbers of disulphide bonds as those present in lipid transfer proteins, chitinases and thaumatin-related proteins [9].

Yeasts producing acid proteases may offer an alternative or supplement to bentonite treatment for removal of undesirable wine proteins [10–13]. In contrast to the classical wine yeast *S. cerevisiae*, non-*Saccharomyces* yeasts are important sources of extracellular enzymes including proteases [6,14]. In the study of Fernández et al. [15], 53 of 141 isolates of "wild yeasts" hydrolyzed casein. The positive strains were identified as *Metschnikowa pulcherrima* and *Pichia membranifaciens*. In a similar study with 245 yeast isolates, 10 strains of *Candida stellata, C. pulcherrima, Kloeckera apiculata* and one strain of *Debaromyces hansenii* showed proteolytic activity [16]. Oenological isolates of *Hanseniaspora* [17], *Metchnikowia pulcherrima* and *Candida apicola* [18] produce extracellular proteases with potential applications in biotechnological processes.

Wine yeasts secreting proteolytic enzymes are of high biotechnological interest for protein haze prevention because they could be directly added as starter cultures to the grape must. Besides cost reductions, there are no administrative restrictions for their applications in must and wine, which should be considered with enzyme preparations.

In this study, we describe the purification and properties of extracellular acid proteases isolated from wine relevant yeasts strains of *Wickerhamomyces anomalus* and *Metschnikovia pulcherrima*, which are produced during cultivation in grape juice.

2. Materials and Methods

2.1. Yeast Strains and Cultures

Yeast strains investigated are deposited at the local culture collection of the Institute for Microbiology and Wine Research (IMW), Johannes Gutenberg University Mainz, Country. The identity of *Wickerhamomyces anomalus* strain 227 and *Metschnikowia pulcherrima* 446 have been verified by sequence analysis of the ITS (internal transcribed spacer region). The primers used for PCR amplification were ITS (F) GGAAGTAAAAGTCGTAACAAGG and ITS (R) TCCTCCGCTTATTGATATGC. Sequencing was performed by LCG Genomics (Berlin, Germany) and identification was accomplished by BLAST searches in public databases.

Yeasts were maintained on GYP medium (20 g·L^{-1} glucose, 10 g·L^{-1} yeast extract, meat peptone 20 g·L^{-1}). Solid media were prepared with 15 g·L^{-1} agar. For cultivation and protease production, a white grape juice (Lindavia®, Niehoffs-Vaihinger Fruchtsaft GmbH, Lauterecken, Germany) was diluted to 50% (v/v) with deionized water and steam-sterilized for 10 min at 100 °C. Volumes of 100 mL or 300 mL of this medium in Erlenmeyer flasks were inoculated with 1% (v/v) of a washed yeast preculture grown in GYP medium. The cultures were incubated on a rotary shaker (100 rpm) at 20 °C for 7 days. On each day, 2 mL samples were taken to monitor cell growth (OD$_{600nm}$) and proteolytic activities in the supernatants.

2.2. Buffer Solutions

Tri-sodium phosphate (20.0 g·L^{-1}, pH 12); sodium phosphate (100 mM, pH 7.0: 17.8 g·L^{-1} NaHPO$_4$ x H$_2$O adjusted with 0.1 M HCl); sodium acetate (100 mM: 6.0 g·L^{-1} glacial acetic acid adjusted to pH 4.5 with 0.1 M NaOH); sodium tartrate (5.0 g·L^{-1} tartaric acid adjusted to pH 3.5 with

0.1 M NaOH); and universal buffer (tris 6.5 g·L^{-1}, maleic acid 2.32 g·L^{-1}, boric acid 1.26 g·L^{-1}; the pH desired was adjusted with 0.1 M NaOH).

2.3. Endo-Protease Activity of Secreted Yeast Enzymes

Endoprotease activities were measured with Megazyme® protease substrate AZCL-collagen® (Megazyme International, Wicklow, Ireland). This compound is prepared by dying and crosslinking collagen to the intensively blue-coloured copper protein azurine to produce a material, which hydrates in water but is still water insoluble. Hydrolysis by proteases produces water-soluble dyed fragments, and the rate of release of these (increase in absorbance at 590$_{nm}$) is directly related to enzyme activity. The standard test was performed as follows: suspensions of AZCL-collagen® were prepared in appropriate buffers depending on the individual experiment. Afterwards, 100 µL of these suspensions were incubated with 100 µL sample for 24 h at 40 °C under vigorous shaking (750 rpm) in a thermo mixer (Eppendorf, Hamburg, Germany). The reaction was stopped by adding 1.0 mL tri-sodium phosphate-buffer (pH 12) and the tubes centrifuged at 16,000 *g* for 30 min. The absorbance of the supernatants were measured at 590$_{nm}$ in a spectrophotometer (Shimadzu UV-2450, Duisburg, Germany). The control solutions were prepared as above, but test samples were heat-treated (30 min, 100 °C) to inactivate any protease activity. Experiments were performed in triplicate.

2.4. Protease Inhibition Studies

Effects of selected inhibitors on proteolytic activities were tested in triplicate by incubation of the samples (1 h) with phenylmethylsulfonylfluoride (PMSF, 2 mM), pepstatin A (20 µM) or Na-EDTA (10 mM) before starting the reaction at 40 °C.

2.5. pH Optimum

The pH optimum of yeast proteases was determined in triplicate with the AZCL-collagen® substrate suspended in universal buffer between pH 3.0 and 8.0. Data presented are mean values of triplicate determinations.

2.6. Temperature Optimum

The temperature optimum of yeast proteases was determined in triplicate with the standard AZCL-collagen® assay at pH 3.5 between 20 and 60 °C.

2.7. Real-Time Monitoring of Proteolytic Yeast Enzymes

A test introduced by Chasseriaud et al. [13] was used with some modifications to monitor proteolytic activity by yeasts directly during growth in grape must. In brief, white grape juice (50% *v/v*) was steam-sterilized for 10 min at 100 °C. A stock solution (20 mg·mL^{-1}) of the chromogenic protease substrate azocasein (Sigma-Aldrich, Munich, Germany) was prepared in 0.1 M NaOH and added to the grape juice at a final concentration of 2.0 mg·mL^{-1}.

At regular intervals, 600 µL samples of the yeast cultures were withdrawn and mixed with 90 µL trichloroacetic acid (20% *w/v*). Samples were centrifuged at 16,000 *g* for 30 min. In addition, 500 µL NaOH (1 M) were added to the supernatants and colour release from azocasein was measured in a spectrophotometer (Shimadzu UV-2450, Duisburg, Germany) at 500$_{nm}$. Non-inoculated culture media served as controls. Experiments were performed in duplicate.

2.8. Preparative Isoelectric Focusing (pIEF)

Purification of the extracellular proteases was achieved by preparative IEF using the Rotofor® Preparative IEF-Cell (BioRad, Munich, Germany). Yeasts were grown for 6 days in 100 mL grape juice on a shaker (100 rpm) at 20 °C. After centrifugation (10,000 *g* for 10 min), supernatants were dialyzed (cut-off: 10 kDa) and lyophilized. Concentrated culture supernatants (3.0–5.0 mL) were filled up to 50 mL with double-deionized water and mixed with 1.0 mL of 40% (w/v) ampholyte solution pH 3.0 to 5.0 (Rotilyte®, Roth, Karlsruhe, Germany). The electrode buffers were 0.5 M acetic acid (anode) and 0.25 M HEPES (cathode), respectively. The separation was performed under a preset power of 15 W until constant voltage was reached after 4 to 5 h at 10 °C. The pH of the 20 fractions obtained was measured, their protein content checked by SDS-PAGE, and protease activities tested with AZCL-collagen®. The protease positive fractions were concentrated in 50 mL spin columns (cut-off: 10 kDa, Vivaspin®, Sartorius AG, Goettingen, Germany), washed three times with deionized water and stored at −20 °C.

2.9. Gel Electrophoretic Methods

Protein compositions in the culture filtrates and fractions of pIEF were checked by SDS-PAGE. Samples were separated in 12.5% (w/v) SDS gels (10 cm × 10 cm) with a 4% (w/v) stacking gel at room temperature for 1 h. A prestained protein ladder (Pink Color Protein Standard II®; Serva, Heidelberg, Germany) served as the molecular mass standard. Protein staining was performed with Quick Coomassie Stain® (Serva, Heidelberg, Germany).

2.10. Identification of Yeast Exoenzymes

The concentrated and dialyzed culture filtrates were separated by SDS-PAGE and gels treated with Quick Coomassie Stain®. *W. anomalus* 227 delivered two well separated protein bands after 24 h incubation, which were excised and sliced into small pieces. After destaining and drying, proteins were reduced with 2 mM DTT at 55 °C and alkylated with 15 mM iodacetamide at room temperature in the dark for 1 h each. After washing and drying, trypsin digests were done at 37 °C overnight (0.5 µg of trypsin per gel slice). Tryptic peptides were transferred into an autosampler vial for peptide analysis via LC-MS/MS. Nanoscale liquid chromatography of tryptic peptides was performed with a Waters NanoAcquity UPLC system (Eschborn, Germany) equipped with a 75 µm × 250 mm HSS-T3 reversed phase column and a 2.6 µL PEEKSIL-sample loop (SGE, Darmstadt, Germany) as described before (PMID: 23265486). Online mass spectrometry analysis of tryptic peptides was performed using a Waters Synapt G2-S QTOF mass spectrometer, operated at a resolving power of $R = 20.000$. All analyses were performed using positive mode ESI using a NanoLockSpray source as described (PMID: 23265486). Resulting liquid chromatography tandem MS (LC-MS/MS) data were processed and searched by using PROTEINLYNX GLOBAL SERVER, Ver. 3.0.2. (Waters, Eschborn, Germany). Protein identifications were assigned by searching a custom compiled database containing open reading frames obtained from transcriptome sequencing of *W. anomalus* [19] for gel bands derived from *W. anomalus* and all available (212.809) RefSeq database entries from the order of *Saccharomycetales* for gel bands derived from *Metschnikowia pulcherrima*. Sequence databases were supplemented by known possible contaminants (trypsin, human keratins) based on the precursor and fragmentation data afforded by the LC-MS/MS acquisition method as described before [20]. The false discovery rate (FDR) for peptide and protein identification was assessed searching a reverse database generated automatically in PLGS. FDR was set to 0.01 for database search.

2.11. In Silico Analysis

Protein sequence was analysed by public databases and tools (BLAST; ScanPROSITE; PROTPARAM; SIGNAL P, NetNGlyc) offered by the ExPASY Bioinformatics Portal.

2.12. Degradation Experiments

Proteins from a German Riesling wine (vintage 2012) were prepared by dialysis and lyophilisation as recently described by Jaeckels et al. [20]. The proteolytic activity against isolated wine proteins and bovine serum albumin (Sigma-Aldrich, Munich, Germany) were tested after incubation at 20 °C in sodium tartrate buffer (pH 3.5) and subsequently in an actual German Riesling wine (vintage 2014,

11.5% v/v alcohol). The specific reaction conditions are indicated in the corresponding results section. Degradation of the model proteins was checked by SDS-PAGE.

3. Results

In a previous screening study with 102 yeasts from the internal strain collection (IMW), *W. anomalus* 227 and *M. pulcherrima* 446 were identified as potential protease producers as evidenced by clearing zones on turbid milk powder agar plates (unpublished). In the present study, both strains grew equally well in 50% grape juice and reached a maximum cell density after six days (Figure 1).

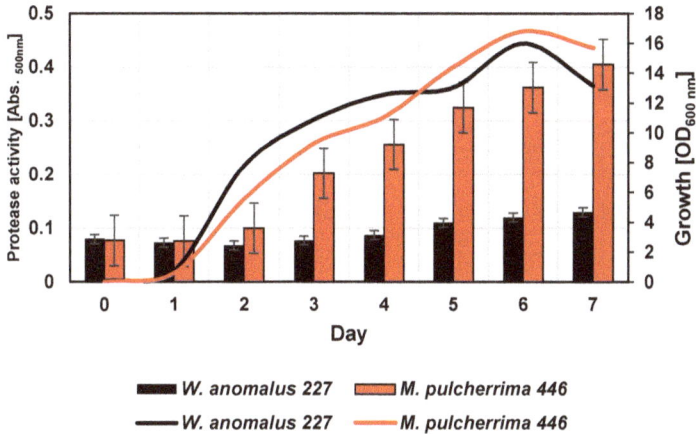

Figure 1. Cell densities (lines) and proteolytic activities (columns) of *Wickerhamoyces anomalus* 227 and *Metschnikovia pulcherrima* 446 during cultivation in a white grape juice at 20 °C. Proteolytic activity was directly detected in the cultures by hydrolysis of supplemented azocasein.

In the course of cultivation, yeasts exhibited proteolytic activities as indicated by cleavage of azocasein supplemented to the medium, although at quite different levels. In order to characterize these activities in more detail, the cultures were harvested after six days and the supernatants were dialyzed and subsequently concentrated by lyophilisation. As detected by SDS-PAGE (Figure 2), only a few distinct proteins were present in the culture concentrates but not in the juice controls (not shown). The protein bands were cut out of the gel and tryptic fragment patterns were analysed by LC-MS/MS. By database research, two extracellular proteins of *W. anomalus* 227 could be identified. The band at ca. 30 kDa corresponded with 100% identity to an exo-β-1,3-glucanase of *W. anomalus* AS1 described in our recent study [21].

The higher molecular mass protein delivered 100% identity with a hypothetic protein (NCBI Acc. No. XP_019036036) annotated in the genome of *W. anomalus* NRRL Y-366-8. It could be attributed to peptidase family A1 (Figure 3). The sequence contains two aspartyl residues (D_{79} and D_{291}) as typically found in the active site of acid proteases. In addition, a putative signal peptide for secretion, possible cysteine bridges and a single N-glycosylation site were detected. Due to the lack of reference sequence data, the extracellular proteins of *M. pulcherrima* 446 were not identified by database searches.

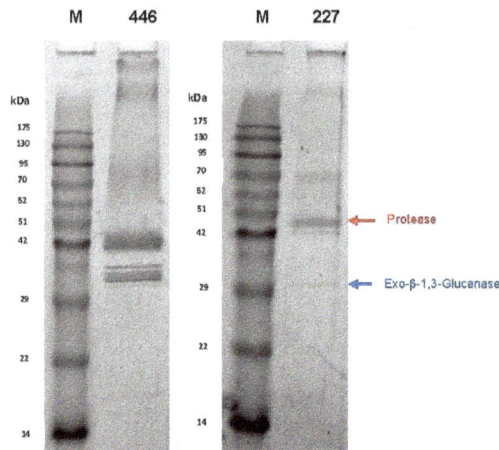

Figure 2. SDS-PAGE (12.5%) of dialyzed and lyophilized culture supernatants after six days of yeast growth in 50% (*v/v*) grape juice. Strains: *M. pulcherrima* 446 and *W. anomalus* 227. M: Molecular mass standard. Arrows indicate proteins identified by LC-MS/MS.

Figure 3. Sequence coverage (peptides identified by LC-MS/MS) and annotated putative protein sequence of the extracellular protease WaAPR1 of *W. anomalus* 227. Amino acids are given in the short code.

Above: Peptides identified by LC-MS/MS are highlighted according to their type (see legend). Total sequence coverage was 49.4%.

Below: Annotated putative protein sequence obtained from the *W. anomalus* transcriptome database. Number of amino acids: 447; Molecular mass: 47328 Da (PROTPARAM); Theoretical pI: 4.11 (PROTPARAM). A possible signal peptide of 19 amino acids (SIGNAL P) is shown in red and underlined. The sequence contains a peptidase family A1 domain (amino acids$_{61-395}$) with aspartyl residues D_{79} and D_{291} (green) in the active site (ScanPROSITE) and a single N-glycoyslation signature (blue; NetGlyc 1.0 server). Cysteine residues C_{329} and C_{260} (yellow) may build a disulfide bridge (ScanPROSITE).

The extracellular proteases of both yeasts were purified by preparative IEF. The proteolytic activity of *W. anomalus* 227 focused at ca. pH 3.4, whereas two activity peaks at pH 3.96 and 4.61 were found in fractions of *M. pulcherrima* 446 (Figure 4).

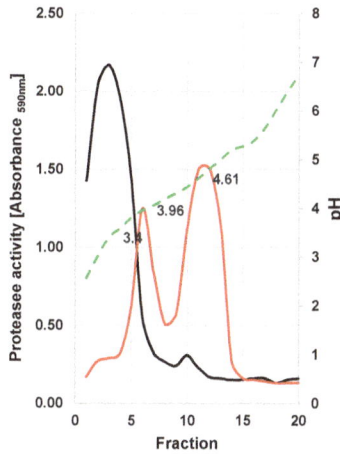

Figure 4. Preparative IEF of concentrated culture supernatants of *W. anomalus* 227 (**black** line) and *M. pulcherrima* 446 (**red** line). Collected fractions were examined for protease activities under the standard assay conditions. The pH gradient is illustrated as the broken **green** line.

The protease-active fractions were pooled, concentrated and subjected to SDS-PAGE and sensitive protein staining (Figure 5). One single band appeared in the fractions collected from the main activity peak of *W. anomalus* 227 (at ≈ pH 3.4) and in the second peak of *M. pulcherrima* 446 (at ≈ pH 4.61).

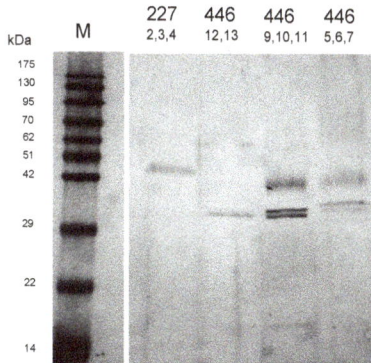

Figure 5. SDS-PAGE (12.5%) of protease-active fractions from *W. anomalus* 227 and *M. pulcherrima* 446 as obtained by preparative IEF.

The isoelectric point and the apparent molecular mass of the purified protease WaAPR1 from *W. anomalus* 227 correspond fairly well with the theoretical values (pI 4.11, MW 47 kDa) derived from the amino acid sequence (Figure 3). The protein purified from the culture supernatant (in fractions 12 and 13) of *M. pulcherrima* 446 is probably one of several protease isoenzymes secreted by this strain. This can be concluded from the appearance of at least five protein bands in SDS-PAGE (Figure 2) and by two activity peaks displayed in preparative IEF. The protein concentrations and enzymatic activities at different stages of purification are listed in Table 1.

Table 1. Protein concentrations and proteolytic activities at different stages of purification.

Species	Sample	Protein Concentration (µg/mL)	Activity * (Absorbance 590 nm)
W. anomalus 227	Original culture supernatant (7 days)	4.87	0.85
	After dialysation and lyophilisation	20.21	1.75
	Concentrated pIEF fractions (2,3,4)	5.85	1.63
M. pulcherrima 446	Original culture supernatant (7 days)	5.27	0.59
	After dialysation and lyophilisation	23.18	2.09
	Concentrated pIEF fractions (12,13)	8.43	1.72

* Determined with Protazyme OL® tablets.

In accordance with the sequence data, inhibitor studies confirmed that both yeasts secrete acid proteases: activities were most significantly impaired by low concentrations of pepstatin A, which specifically inhibits aspartic proteases (Table 2).

Table 2. Effect of inhibitors on protease activities of *W. anomalus* 227 and *M. pulcherrima* 446.

Inhibitor	Concentration	Activity (%)*	
		227	446
Pepstatin A	20 µM	18.7	7.4
PMFS	2 mM	58.7	47.1
EDTA	10 mM	49.6	44.0

* Relative to the control without inhibitor. Activity was determined with Protazyme OL® tablets.

Inhibitors for serine proteases (PMFS) and metalloproteinase (EDTA) were far less effective even at high concentrations. In addition, enzymatic activity on azurine collagen was highest at pH 3–4 and dropped at higher pH values (Figure 6).

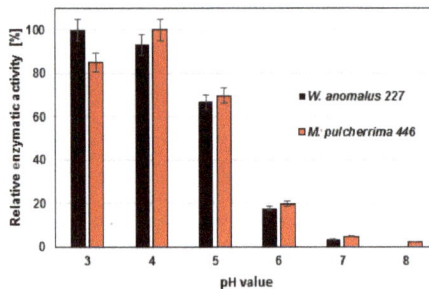

Figure 6. Effect of pH on protease activities of *W. anomalus* 227 and *M. pulcherrima* 446.

The temperature maximum of proteolytic activity for both yeast strains was found at 40 °C, although significant activities could be determined at lower and higher temperatures (Figure 7).

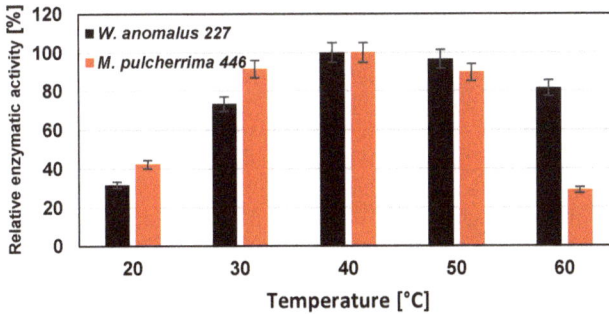

Figure 7. Effect of temperature on protease activities of *W. anomalus* 227 and *M. pulcherrima* 446.

With respect to a potential application for protein haze reduction, we finally assessed the activity at wine-relevant conditions. Bovine serum albumin was completely degraded after 24 h in a white wine. However, thaumatin-like wine proteins were resistant to enzymatic hydrolysis (Figure 8).

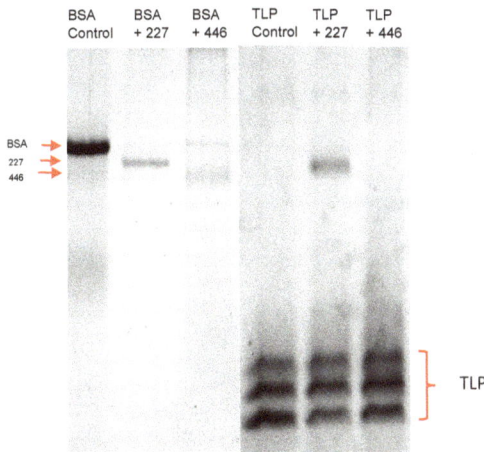

Figure 8. Detection of proteolytic activity of yeasts by SDS-PAGE (12.5%). Bovine serum albumine (BSA, **left**) and thaumatin-like wine proteins (TLP, **right**) were added to a white wine at final concentrations of 100 μg·mL^{-1} and incubated for 24 h at 20 °C with concentrated culture supernatants of *W. anomalus* 227 and *M. pulcherrima* 448.

4. Discussion

Aspartic proteases are ubiquitously distributed proteolytic enzymes that are active in acidic environments [22]. Several microorganisms secrete such proteases as virulence factors and/or in order to break down proteins, thereby deliberating sources of nitrogen.

Extracellular proteinases secreted by yeasts have been isolated and examined by several researchers, especially with respect to their enzymatic properties and the physiology of their induction and secretion. Most of these studies focus on the acid aspartic proteases of *Candida albicans* because of their involvement in human diseases. Nowadays, there is growing interest on yeast acid proteases for industrial applications [10].

Although simple screening procedures with solid casein agar plates have repeatedly demonstrated proteolytic activities in various *non-Saccharomyces* species, only limited knowledge exists on their cellular location, substrate preferences, catalytic properties or molecular structure [11,15,16]. In contrast to some previous studies using synthetic culture media and proteins as inducers [18,23], we demonstrated the secretion of proteases by two wine associated yeasts directly in a natural substrate, i.e., grape juice. Remarkably, they were the only dominant extracellular enzymes produced during the active growth phase. Recent studies have already demonstrated protease activities in *Wickerhamomyces* isolates from enological ecosystems [24]. However, the activities detected were relatively low and not expressed in all strains examined [25]. The extracellular protease WaAPR1 of *W. anomalus* 227 with a molecular mass of 47 kDa and an isoelectric point of 4.11 is in the typical range of microbial aspartic proteases [10,22]. As only a single N-glycosylation site is present in the protein sequence, any covalent glycosylation would be difficult to demonstrate. A similar situation has been reported for the aspartic protease from *M. pulcherrima* IWBT Y1123 [18]. The bioinformatic tool ScanPROSITE predicts the position of the two essential aspartyl residues in the active site at D_{79} and D_{291}. Results of a BLAST search reveals 100% identity with a hypothetic protein (NCBI Acc. No. XP_019036036) in the genome of *W. anomalus* NRRL Y-366-8 and ca. 35% identity with putative acid proteases of other yeast species belonging to the genera *Wickerhamomyces, Saccharomyces, Zygosaccharomyces, Metschnikovia, Yarrowia, Torulaspora* and *Candida*.

The amino acid sequence of the *M. pulcherrima* 446 protease could not be identified in this study. This is attributed to a lack of reference sequence data. In addition, the identification attempts reported in our study were performed with a non-purified concentrate (Figure 2), which probably contained a mixture of several protease isoenzymes. The MALDI analysis will be repeated with a purified sample. Nevertheless, biochemical data (pH optimum, inhibition by pepstatin A) and results of other researchers give strong hints for an aspartic protease. Reid et al. [18] reported on genetic and biochemical levels of the existence of an extracellular aspartic protease in *M. pulcherrima* strain IWBT Y1123. Gotoh et al. [23] purified a protease from the culture supernatant from strain KSY 188-5. The latter had an apparent molecular mass of 37 kDa, an isolelectric point of 4.7, and an optimum activity around 45 °C at pH 3.0. These parameters are very close to the data obtained in the present study for the protease of *M. pulcherrima* 446.

Constitutive production in the natural substrate grape juice without need of external inducers combined with maximum activity at low pH suggests an essential physiological role of the secreted enzyme. The protease activity might help to gain nitrogen from grape proteins necessary for cell growth, especially considering the surplus of sugar carbon in nitrogen-limited milieus [26,27]. The secreted proteases may also act as a survival tool by degrading cell wall proteins of competitive microorganisms, analogous to its role as a virulence factor in *C. albicans*. The membrane-bound aspartic protease of *S. cerevisiae*, a member of the yapsin protease family, is involved in cell wall growth and maintenance. It is a challenge to get more insights in the natural role of secreted yeast aspartic proteases. On the other hand, their biotechnological potential needs to be further examined in future experiments. The use of aspartic proteases as alternatives or supplements to clarifying agents in various beverage industries is under intensive investigation, and the potential applications in the wine industry are thoroughly discussed [10]. Under wine-relevant conditions (pH 3.5, 20 °C), the proteases of both yeast strains were able to degrade bovine serum albumin but not wine specific thaumatin-like proteins (TLP). Nevertheless, it is well known that TLP are very inert to proteolytic hydrolysis and among the few protein classes that finally reach the bottled wine. One should keep in mind that even the actual used commercial fungal protease preparation is only effective at elevated (not wine-relevant) temperatures and cannot completely eliminate wine proteins. Other grape must proteins than TLP may be more susceptible to proteolytic hydrolysis by yeast enzymes. The fact that the proteolytic enzymes were produced during growth in grape juice offers the chance to lower the protein content by using our yeast strains directly for must fermentations. Recent studies have already shown that strains of *Wickerhamomyces anomalus* and *Metschnikovia pulcherrima* positively influence wine sensory properties [28]. Any reduction of grape must protein content would mean an advantage for wine-makers by lowering the effective bentonite dosages. These aspects are topics of our current research.

Acknowledgments: The authors would like to thank undergraduate student Hendrik Leibhan for conducting some of the experiments.

Author Contributions: Martina Schlander performed most of the experiments; Ute Distler performed HPLC-MS analysis; Stefan Tenzer supervised and evaluated the HPLC-MS/MS analysis; Eckhard Thines provided financial, laboratory and personnel capacities; Harald Claus conceived and designed the experiments, analyzed the data and wrote the manuscript.

Conflicts of Interest: The authors declare no conflict of interest.

Abbreviations

TLP   Thaumatin-like Proteins
pIEF   Preparative Isoelectric Focusing

References

1. Padilla, B.; Gil, J.V.; Manzanares, P. Past and future of Non-*Saccharomyces* yeasts: From spoilage microorganisms to biotechnological tools for improving wine aroma complexity. *Front. Microbiol.* **2016**. [CrossRef] [PubMed]
2. Ciani, M.; Morales, P.; Comitini, F.; Tronchoni, J.; Canonico, L.; Curiel, J.A.; Oro, L.; Rodrigues, A.J.; Gonzalez, R. Non-conventional yeast species for lowering ethanol content of wines. *Front. Microbiol.* **2016**. [CrossRef] [PubMed]
3. Maturano, Y.P.; Assof, M.; Fabani, M.P.; Nally, M.C.; Jofré, V.; Rodríguez Assaf, L.A.; Toro, M.E.; Castellanos de Figueroa, L.I.; Vazquez, F. Enzymatic activities produced by mixed *Saccharomyces* and non-*Saccharomyces* cultures: Relationship with wine voltile composition. *Antonie van Leeuwenhoek* **2015**, *108*, 1239–1256. [CrossRef] [PubMed]
4. Suzzi, G.; Schirone, M.; Sergi, M.; Marianella, R.M.; Fasoli, G.; Aguzzi, I.; Tofalo, R. Multistarter from organic viticulture for red wine Montepulciano d'Abruzzo production. *Front. Microbiol.* **2012**. [CrossRef]
5. Tofalo, R.; Patrignani, F.; Lanciotti, R.; Perpetuini, G.; Schirone, M.; Di Gianvito, P.; Pizzoni, D.; Arfelli, G.; Suzzi, G. Aroma profile of Montepulciano d'Abruzzo wine fermented by single and co-culture starters of autochthonous *Saccharomyces* and non-*Saccharomyces* yeasts. **2016**. [CrossRef]
6. Claus, H. Exoenzymes of wine microoorganisms. In *Biology of Microorganisms on Grapes, in Must and Wine*; König, H., Unden, G., Fröhlich, J., Eds.; Springer: Berlin/Heidelberg, Germany, 2009; pp. 259–271.
7. Mateo, J.J.; Maicas, S. Application of Non-*Saccharomyces* yeasts to wine-making process. *Fermentation* **2016**, *2*, 14. [CrossRef]
8. Ugliano, M. Enzymes in winemaking. In *Wine Chemistry and Biochemistry*; Moreana-Arribas, M.V., Polo, C., Eds.; Springer Science-Business Media: Adelaide, Australia, 2009; pp. 103–126.
9. Van Sluyter, S.C.; McRae, J.M.; Falconer, R.J.; Smith, P.A.; Bacic, A.; Waters, E.J.; Marangon, M. Wine protein haze: Mechanisms of formation and advances in prevention. *J. Agric. Food. Chem.* **2015**, *63*, 4020–4030. [CrossRef] [PubMed]
10. Theron, L.W.; Divol, B. Microbial aspartic proteases: Current and potential applications in industry. *Appl. Microbiol. Biotechnol.* **2014**, *98*, 8853–8863. [CrossRef] [PubMed]
11. Rosi, I.; Costamagna, L.; Bertuccioli, M. Screening for extracellular protease(s) production by wine yeasts. *J. Inst. Brew.* **1987**, *93*, 322–324. [CrossRef]
12. Lagace, L.S.; Bisson, L.F. Survey of yeast acid proteases for effectiveness of wine haze reduction. *Am. J. Enol. Vitic.* **1990**, *41*, 147–162.
13. Chasseriaud, L.; Miot-Sertier, C.; Coulon, J.; Iturmendi, N.; Moine, V.; Albertin, W.; Bely, M. A new method for monitoring the extracellular proteolytic activity of wine yeasts during alcoholic fermentation of grape must. *J. Microbiol. Methods* **2015**, *119*, 176–179. [CrossRef] [PubMed]

14. Molnárova, J.; Vadkertiová, R.; Stratilová, E. Extracellular enzymatic activities and physiological profiles of yeasts colonizing fruit trees. *J. Basic Microbiol.* **2014**, *51*, S74–S84. [CrossRef] [PubMed]
15. Fernández, M.; Úbeda, J.F.; Briones, A.I. Typing of non-*Saccharomyces* yeasts with enzymatic activities of interest in wine-making. *Int. J. Food. Microbiol.* **2000**, *59*, 29–36. [CrossRef]
16. Strauss, M.L.A.; Jolly, N.P.; Lambrechts, M.G.; van Rensburg, P. Screening for the production of extracellular hydrolytic enzymes by non-*Saccharomyces* wine yeasts. *J. Appl. Microbiol.* **2001**, *91*, 182–190. [CrossRef] [PubMed]
17. Mateo, J.J.; Maicas, S.; Thießen, C. Biotechnological characterisation of extracellular proteases produced by enological *Hanseniaspora* isolates. *Int. J. Food Sci. Technol.* **2015**, *50*, 218–225. [CrossRef]
18. Reid, V.J.; Theron, L.W.; du Toit, M.; Divol, B. Identification and partial characterization of extracellular aspartic protease genes from *Metschnikowia pulcherrima* IWBT Y1123 and *Candida apicola* IWBT Y1384. *Appl. Environm. Microbiol.* **2012**, *78*, 6838–6849. [CrossRef] [PubMed]
19. Schneider, J.; Fupp, O.; Trost, S.; Jaenicke, S.; Passoth, V.; Goesmann, A.; Tauch, A.; Brinkrolf, K. Genome sequence of *Wickerhamomyces anomalus* DSM 6766 reveals genetic basis of biotechnologically important antimicrobial activities. *FEMS Yeast Res.* **2012**, *12*, 382–386. [CrossRef] [PubMed]
20. Jaeckels, N.; Tenzer, S.; Rosfa, S.; Schild, H.; Decker, H.; Wigand, P. Purification and structural characterization of lipid transfer protein from red wine and grapes. *Food Chem.* **2013**, *138*, 263–269. [CrossRef] [PubMed]
21. Schwentke, J.; Sabel, A.; Petri, A.; König, H.; Claus, H. The wine yeast *Wickerhamomyces anomalus* AS1 secretes a multifunctional exo-β-1,3 glucanase with implications for winemaking. *Yeast* **2014**, *31*, 349–359. [CrossRef] [PubMed]
22. Ogrydziak, D.M. Yeast extracellular proteases. *Crit. Rev. Biotechnol.* **1993**, *13*, 1–55. [CrossRef] [PubMed]
23. Gotoh, T.; Kikuchi, K.; Kodama, K.; Konno, H.; Kakuta, T.; Koizumi, T.; Nojiro, K. Purification and properties of extracellular carboxyl proteinase secreted by *Candida pulcherrima*. *Biosci. Biotech. Biochem.* **1995**, *59*, 367–371. [CrossRef]
24. Madrigal, T.; Maicas, S.; Mateo Tolosa, J.J. Glucose and ethanol tolerant enzymes produced by *Pichia* (*Wickerhamomyces*) isolates from enological ecosystems. *Am. J. Enol. Vitic.* **2013**, *64*, 126–133. [CrossRef]
25. López, S.; Mateo, J.J.; Maicas, S. Screening of *Hanseniaspora* strains for production of enzymes with potential interest for winemaking. *Fermentation* **2016**. [CrossRef]
26. Christ, E.; Kowalczyk, M.; Zuchowska, M.; Claus, H.; Löwenstein, R.; Szopinska-Morawska, A.; Renaut, J.; König, H. An exemplary model study for overcoming stuck fermentation during spontaneous fermentation with the aid of a *Saccharomyces* triple hybrid. *J. Agric. Sci.* **2015**, *7*, 18–34. [CrossRef]
27. Guo, Z.P.; Zhang, L.; Ding, Z.Y.; Wang, Z.X.; Shi, G.Y. Improving the performance of industrial ethanol-producing yeast by expressing the aspartyl protease on the cell surface. *Yeast* **2010**, *27*, 1017–1027. [CrossRef] [PubMed]
28. Varela, C. The impact of non-*Saccharomyces* yeasts in the production of alcoholic beverages. *Appl. Microbiol. Biotechnol.* **2016**, *100*, 9861–9874. [CrossRef] [PubMed]

fermentation

MDPI

Article

Gravity-Driven Adaptive Evolution of an Industrial Brewer's Yeast Strain towards a Snowflake Phenotype in a 3D-Printed Mini Tower Fermentor

Andreas Conjaerts and Ronnie G. Willaert *

Alliance Research Group VUB-UGent NanoMicrobiology (NAMI), IJRG VUB-EPFL,
BioNanotechnology & NanoMedicine (NANO), Structural Biology Brussels (SBB),
Vrije Universiteit Brussel, Brussels 1050, Belgium; Andreas.Conjaerts@vub.ac.be
* Correspondence: Ronnie.Willaert@vub.ac.be; Tel.: +32-2-629-1846

Academic Editors: Maurizio Ciani and Badal C. Saha
Received: 29 August 2016; Accepted: 3 January 2017; Published: 5 January 2017

Abstract: We designed a mini tower fermentor that is suitable to perform adaptive laboratory evolution (ALE) with gravity imposed as selective pressure, and suitable to evolve a weak flocculating industrial brewers' strain towards a strain with a more extended aggregation phenotype. This phenotype is of particular interest in the brewing industry, since it simplifies yeast removal at the end of the fermentation, and many industrial strains are still not sufficiently flocculent. The flow of particles (yeast cells and flocs) was simulated, and the theoretical retainment advantage of aggregating cells over single cells in the tower fermentor was demonstrated. A desktop stereolithography (SLA) printer was used to construct the mini reactor from transparent methacrylic acid esters resin. The printed structures were biocompatible for yeast growth, and could be sterilised by autoclaving. The flexibility of 3D printing allowed the design to be optimized quickly. During the ALE experiment, yeast flocs were observed within two weeks after the start of the continuous cultivation. The flocs showed a "snowflake" morphology, and were not the result of flocculin interactions, but probably the result of (a) mutation(s) in gene(s) that are involved in the mother/daughter separation process.

Keywords: 3D printing; mini tower fermentor; *Saccharomyces cerevisiae*; industrial brewer's yeast; adaptive laboratory evolution (ALE); snowflake phenotype

1. Introduction

The use of evolutionary methods is a more "natural" approach to enhance the attributes of microorganisms, in contrast to genetic modification which has so far precluded its commercial use due to the low consumer tolerance for genetically modified organisms [1,2]. Evolutionary methods can be applied even before the genetic elements and their global interactions required for optimal performance by an organism are understood. Recently, artificial laboratory selection of microbial cells has been introduced to generate potentially robust and optimised microbial production systems [3,4]. Adaptive laboratory evolution (ALE) strategies allow for the metabolic engineering of microorganisms by combining genetic variation with the selection of beneficial mutations in an unbiased fashion [5]. A number of investigations have demonstrated the feasibility of directing evolution in natural *Saccharomyces pastorianus* hybrid stains in order to create variant strains with improved functional properties [6]. Such investigations have focused on adaptation to very high-gravity brewing conditions [7–9], associated stresses (such as osmotic stress and ethanol toxicity) [10,11], or the modification of the production of flavour compounds [12]. ALE has also been utilised to enhance the fermentation rate of *S. cerevisiae* with decreased formation of acetate and greater production of aroma compounds [13,14].

Continuous culture provides many benefits over classical batch-style cultivation to perform experimental evolution [15]. Steady-state cultures allow for precise control of growth rate and environment, and cultures can be propagated for weeks or months in these controlled environments, which is important for the study of experimental evolution. Continuous mini bioreactors have been successfully used as multiplexed chemostat arrays for adaptive evolution experiments with yeast cells [15,16]. The use of mini bioreactors has several advantages, such as reduced costs for media and labour, and the ability to perform a large number of fermentations in parallel [17–19].

Recently, three-dimensional (3D) printing has been used for medical applications [20], such as (1) medical models; (2) medical aids, orthoses, splints, and prostheses; (3) tools, instruments, and parts for medical devices; (4) inert implants; and (5) biomanufacturing [21–24]. Bioprocess applications include tissue engineering scaffolds and corresponding bioreactors [25]. 3D printing has recently been used to fabricate fullerene-type biocarriers for biofilm growth that can be used in bioreactors for wastewater treatment [26]. 3D printing technology shows great potential for the easy development of mini-scale bioreactors. Here, we report the use of 3D printing for the construction of a continuous mini tower fermentor.

The concept of 3D printing—also referred to as additive manufacturing, rapid prototyping, or solid-freeform fabrication (SFF)—was developed by Charles Hull in the early 1980's [27]. 3D printing is used for rapid prototyping of 3D models originally generated by a computer aided design (CAD) program. The 3D model is sliced into 2D horizontal cross sections, which are printed in consecutive layers. There are several well-established methods of 3D-printing, such as stereolithography (SLA), fused deposition modelling, selective laser sintering, multi jet fusion, and selective laser melting. SLA became the first commercialised 3D-printing technique (invented by Charles Hull in 1983), and remains one of the most powerful and versatile of all SFF techniques [28]. SLA works by exposing a layer of photosensitive liquid resin to a UV-laser beam so that the resin hardens and becomes solid. Once the laser has swept a layer of resin in the desired pattern and it begins to harden, the model-building platform in the liquid tank of the printer steps down the thickness of a single layer, and the laser begins to form the next layer. SLA is capable of printing at high resolutions (up to 25 μm) and relatively high production rates (1.5 cm/h).

In this contribution, we designed a continuous mini bioreactor that is suitable to perform ALE of a weak flocculating industrial brewers' strain towards a strain with a more extended aggregation (flocculation) phenotype. Yeast strains that aggregate or flocculate are of particular interest to the brewer, since it simplifies yeast removal at the end of the primary fermentation [29]. Since many industrial brewing strains still show no adequate flocculation, we applied ALE as a non-genetic engineering method. The design of the fermentor was based on a continuous mini tower fermentor; i.e., the continuous A.P.V. tower fermentor that was used in the 1960s to produce beer on an industrial scale in a British brewery [30,31]. In this type of fermentor, gravity is the selective pressure to enhance the aggregation phenotype during evolution. We demonstrate that SLA 3D printing can be used to construct the mini tower fermentor, and that the design is suitable to obtain a yeast cell aggregation ("snowflake") phenotype by performing ALE.

2. Materials and Methods

2.1. Yeast Strains and Media

An industrial *S. cerevisiae* brewer's strain was used; i.e., CMBSVM22 from the yeast collection of prof. em. Freddy Delvaux (Centre for Malting and Brewing Science (CMBS), Katholieke Universiteit Leuven, Leuven Belgium; Biercentrum Delvaux, Beekstraat 20, 3040 Neerijse, Belgium) [32]. The haploid *S. cerevisiae* BY4742 strain [33] was used for the biocompatibility assay, and the flocculent BY4742 [*FLO1*] strain [34] for the preliminary experiments during experimental design optimisation. These strains were precultured in YPD (Yeast extract-Peptone-Dextrose) medium (1% *m/v* yeast extract, 2% *m/v* meat peptone, 4% *m/v* D-glucose) overnight at 30 °C. Growth medium (100 g/L D-glucose,

4 g/L $(NH_4)SO_4$, 1.5 g/L KH_2PO_4, 1 g/L $MgSO_4.7H_2O$, and 5 g/L yeast extract) was used for the continuous adaptive evolution experiment in the mini tower bioreactor.

WL (Wallerstein Laboratory) nutrient agar was used to differentiate between colonies by morphology variations [35,36]. The medium contained 7.5 g WL nutrient agar per 100 mL deionized water. The cell suspensions were diluted to 10×10^6 cells/mL with sterile deionized water. Four droplets of 1 µL of the diluted suspension were spotted on each plate.

2.2. Determination of Cell Concentration and Growth Rate

The cell concentration was determined by cell dry mass (CDM) and optical density (OD) measurements. For CDM measurements, a sampled cell suspension of 3 mL was filtered with a 0.45-µm filter (Type HVLP, Millipore®, Darmstadt, Germany), and the filter was washed with deionised water. The filter was dried at 80 °C until the mass remained stable. The difference in mass between the filter before and after filtration gave the cell dry mass (CDM) concentration (g·CDM/L). For the OD measurements, absorption of the cell suspension was determined at a wavelength of 600 nm. Biomass measurements were performed in triplicate, and the standard deviation was calculated to represent the error bars. First-order growth kinetics was used to fit the exponential part of the growth curve and to determine the maximum specific growth rate.

2.3. Flocculation Assay

The assay was based on the method of D'Hautcourt and Smart [37], with slight modifications. Cells were grown for 24 h in 20 mL YPD. The cell suspension was centrifuged (4000 rpm, 3 min), and the medium was discarded. Subsequently, the cells were resuspended in an equal volume of EDTA-buffer to chelate the Ca^{2+}-ions and deflocculate the yeast flocs. The $OD_{600\,nm}$-value of the suspension was determined, and such a volume was transferred to an Eppendorf tube to ensure an $OD_{600\,nm}$-value of 10 in 1 mL. EDTA (ethylenediaminetetraacetic acid)-buffer (50 mM EDTA, pH 7) was added until a final volume of 1 mL was acquired. A sample of 50 µL was taken 0.5 mL below the meniscus, and the sample was diluted 20 times in a 1.5 mL cuvette with EDTA-buffer. The tubes were centrifuged (4000 rpm, 3 min), and the supernatant was discarded. The cells were resuspended in 1 mL flocculation buffer A (3 mM $CaSO_4$). The last step was repeated, but the cells were resuspended in flocculation buffer B (3 mM $CaSO_4$, 83 mM CH_3COONa, 4% v/v ethanol, pH 4.5). The tubes were shaken horizontally at 100 rpm for 10 min. Prior to taking 50 µL samples 0.5 mL below the meniscus, 3 min of sedimentation in a vertical position took place. The sample was diluted 20 times with EDTA-buffer in a 1.5 mL cuvette. The absorbance of both suspensions in the cuvettes was determined, and the related flocculation percentage was calculated:

$$\text{Flocculation percentage (\%)} = \frac{OD_{EDTA} - OD_{Flocculation\,buffer}}{OD_{EDTA}} \times 100$$

with OD_{EDTA} as the $OD_{600\,nm}$-value of cells in EDTA-buffer, and $OD_{Flocculation\,buffer}$ the $OD_{600\,nm}$-value of cells in flocculation buffer B.

2.4. Biocompatibility Assay

A commercial resin (Clear Resin GPCL02, FormLabs, Somerville, MA, USA) composed of a mixture of methacrylic acid esters and photoinitiators was used to print the mini reactor. To ensure that this resin had no negative effect on the growth of yeast cells, a biocompatibility assay was performed. Therefore, four sterile Erlenmeyer flasks were filled with 50 mL YPD. Prior to adding a piece of resin to two of the four flasks, the pieces were rinsed with soap water, ethanol, and sterile water. Afterwards, all flasks were inoculated with 4 mL of a preculture of the *S. cerevisiae* BY4742 strain in such a way that a final $OD_{600\,nm}$-value of 0.2 was reached. The flasks were incubated at 30 °C and shaken at 120 rpm. Every two hours, a 1 mL sample was taken from each flask, and the cell concentration was determined.

2.5. Mini Tower Fermentor Design and Construction

The mini tower reactor was designed in COMSOL Multiphysics 4.4 (Stockholm, Sweden). The bioreactor had a high-aspect-ratio (beer depth divided by the inner diameter) of around 8:1, which is based on the APV tower fermentor that was used to produce lager and ale beer in the 1960s [30,31,38,39]. The inlet tube had an inner radius of 1 mm and a length of 18 mm; the outlet tube had an inner radius of 2 mm and a length of 25 mm; both contained barbs to facilitate the attachment of the silicone tubing. The bottom of the vessel was conical with a bottom inner radius of 0.7 mm, a top radius of 5.5 mm, and a wall thickness of 1 mm (Figure 1F). The mid cylindrical part was 88 mm high and had an inner radius of 5.5 mm. Another cone connected the cylinder to the reactor head. This cone had a bottom inner radius of 5.5 mm, a top inner radius of 11 mm, and a height of 11 mm. A conical opening was incorporated in the reactor head with a bottom radius of 7.5 mm, a top inner radius of 9 mm, and a height of 20 mm. The dimensions of this opening fitted a silicone stopper that was pierced with a hypodermic needle (sharp 18G, 150 mm, Samco, Surrey, UK), allowing the reactor to be aerated at the bottom cone.

The benchtop SLA 3D printer Form 1+ (Formlabs, Somerville, MA, USA) was used to construct the bioreactor. The COMSOL Multiphysics design was exported as an "stl" file and imported in the printing software PreForm 2 (Formlabs, Somerville, MA, USA). A layer thickness of 0.05 mm was selected for the printing job. Supports settings were: a density of 1, point size of 0.6 mm, and selection of no internal supports. The reactor was oriented with the top of the reactor towards the base, under an angle of 20° around the z-axis. The printed structure was shaken gently in two consecutive isopropanol baths, and was sterilised by autoclaving (121 °C, 20 min).

In preliminary experiments, the design was optimised by performing short-term cultivations with a strong flocculating yeast strain (BY4742 [*FLO1*]). The design adaptations are illustrated in Figure 1A–F. The medium was fed through the bottom inlet, and was disposed through the middle outlet. To prevent flocs from sedimenting through the inlet tube, the inner radius of the inlet was decreased to obtain a higher medium inlet velocity (Figure 1B–F versus A). In the initial designs (Figure 1A–D), a gas outlet was present at the top of the reactor. To enhance the medium flow through the reactor, this outlet was removed in the final design, and gases were removed via the liquid outlet. The inlet and outlet were modified by printing of barbs at the outsides, which allowed the silicone tubing to be attached strongly using plastic straps (Figure 1G). The bottom of the vessel was conical, which facilitated sedimentation and fluidisation of the yeast flocs. The body of the reactor was a long cylinder. The wider cylindrical reactor head was connected to the body by another cone to avoid floc retention within the reactor head. Therefore, the length and slope of this cone was increased from the second design on. A conical opening was incorporated in the reactor head and fitted with a silicone stopper. This allowed the reactor to be gently aerated and fluidised from the bottom by piercing the silicone stopper with a long needle.

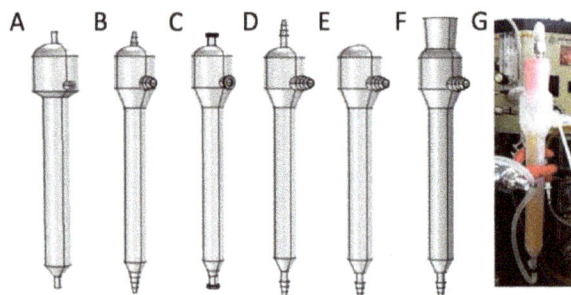

Figure 1. The chronological stages in the development of the 3D-printed mini tower fermentor (**A–F**); (**G**) the printed reactor **F** setup during the evolution experiment.

2.6. Mini Tower Fermentor Mathematical Modelling

To evaluate the effect of gravity as selective pressure within the mini tower reactor, fluid flow and particle tracing simulations were performed in COMSOL Multiphysics 4.4 by selecting the Fluid Flow (Single-Phase Flow, Laminar Flow) and the Particle Tracing for Fluid Flow physics. Water at 20 °C was selected as the working liquid (density 999.62 kg/m^3, dynamic viscosity 0.001 Pa·s). Laminar fluid flow through the reactor was simulated with laminar inflow as boundary condition at the inlet. An inlet flow rate of 1.47 mL/h (with an entrance length of 5 cm) was employed, which was in agreement with the initial flow rate of the actual experiment (see Section 2.7). Pressure (atmospheric pressure) was selected as the boundary condition at the outlet, and backflow was suppressed. Physics-controlled mesh was selected as mesh settings with normal element size. Particle Tracing for Fluid flow was used to calculate the motion of the particles in a background laminar fluid flow. Three thousand particles were released at the inlet, with a random diameter between 3 μm and 70 μm. This range includes the size of planktonic cells as well as small yeast flocs. The density of the particles was set at 1117 kg/m^3, which is the average of values found in the literature [40–43]. The particles were released at the inlet (bottom) of the reactor. The forces acting on the particles are the drag force of the fluid and the gravity force.

2.7. Experimental Set-Up

A fermentation vessel (5 L) (BioFlo III, New Brunswick Scientific, Edison, NJ, USA), which was aerated, agitated at 137 rpm, and heated to 32 °C, was used as medium reservoir, and connected to the mini tower fermentor inlet with silicone tubing (Figure 1G). The peristaltic pump of the BioFlo III fermentor was used as feeding pump. This pump was initially set at a flow rate of 1.47 mL/h, which corresponds to a dilution rate of 0.11 h^{-1}. A silicone stopper was inserted in the mini reactor head and was pierced by a needle to reach the bottom of the reactor. The needle was connected to an aquarium pump to aerate the reactor. The outlet of the reactor was connected to a four-port manifold. Each outlet of the manifold was connected to a needle head. This needle was pierced through a silicone stopper, which was put in a 15- or 50-mL Falcon tube. The manifold enabled fast and simple exchanging of the outlet reservoir. The four Falcon tubes were permanently kept on ice. Inoculation was performed by piercing a syringe through the silicone stopper and ejecting the inoculum directly into the mini reactor.

3. Results

3.1. Simulation of Single Cell and Cell Aggregate Behaviour in the Mini Tower Fermentor

A tower fermentor is suitable to exploit gravity as selective pressure in evolution experiments, since aggregates will be retained preferentially in the reactor compared to single cells. The selective pressure within the tower fermentor was evaluated by particle tracing simulations in COMSOL Multiphysics. Only particles with a diameter smaller than approximately 11 μm were able to reach the top of the reactor (Figure 2). This corresponds to single cells or possibly a complex of two or three cells. According to the calculations, it is very unlikely for larger particles (such as flocs) to reach the outlet of the reactor. Over a longer period of time, this would result in an amplification of flocculating cells within the fermentor. Cells are also retained at the locations of the conical structures. These simulations do not take into account the upward air flow or CO_2 release by the cells during cultivation, which will increase the flow through the reactor. As experimentally demonstrated (see further), these latter conditions also result in a higher probability of cell aggregate retention in the reactor.

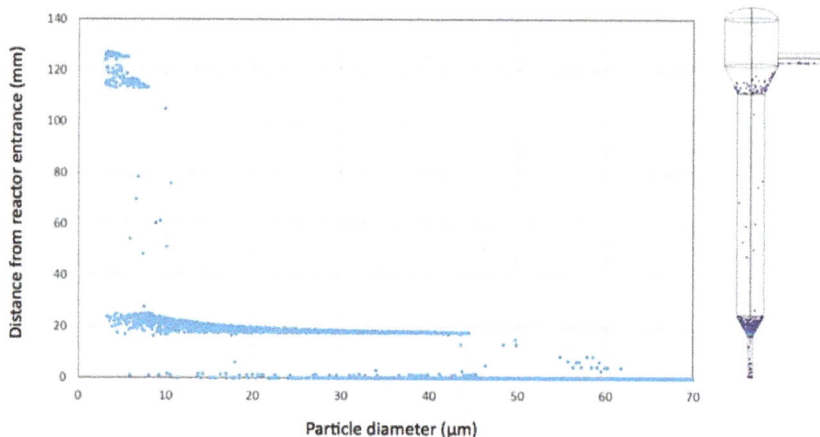

Figure 2. Particle tracing in a 3D-printed bioreactor using COMSOL Multiphysics. The distance travelled from the reactor entrance is plotted against the cell particle diameter.

3.2. Biocompatibility Assay

To ensure that the used resin for the construction of the reactor did not have any negative effects on the growth of the obtained yeast strains, a biocompatibility assay was performed. The *S. cerevisiae* BY4742 strain was grown in the presence and absence of a printed structure. The printed resin material had no effect on the growth of the yeast cells (see Appendix A: Figure A1). The maximum specific growth rate in the presence and absence of the resin material were comparable (0.21 h^{-1} and 0.22 h^{-1}, respectively), indicating the yeast growth biocompatibility of the polymeric material.

3.3. Adaptive Evolution during Continuous Operation

The mini tower fermentor was used for an evolution experiment with the intention of obtaining a more flocculent phenotype of an industrial brewer's strain. The strain was characterised by a very low flocculation percentage (4% ± 6%). The conditions inside the tower fermentor were monitored by sampling at the exit, and by measuring the pH and cell density (Figure 3). The cultivation was initiated with a dilution rate of approximately 0.1 h^{-1}. The cell concentration visibly increased inside the reactor over time. The dilution rate was increased in steps to avoid wash-out and selectively retain larger cell aggregates. The dilution rate was doubled after seven days, which resulted in an increased biomass concentration. The dilution rate was further increased up to 0.4 h^{-1}. At this rate, a dense yeast slurry containing cell aggregates was present in the reactor. The experiment was ended after 14 days of continuous culture when a dense yeast slurry could be visually observed in the reactor.

The yeast population was subsequently examined by optical microscopy (Figure 4). The presence of multicellular yeast flocs was observed. The morphology of the flocs can be described as "snowflake" aggregates. Yeast cell clusters with a comparable morphology have been recently described [44–46]. The evolved cells were plated out on WL nutrient agar, which can be used to detect morphological differences between strains [36]. The population at the start, and after seven and fifteen days were compared (Figure 5). The CMBSVM22 strain possesses the *FLO11* gene required for invasive growth [32,47–49], which could clearly be seen on the bottom of the plates. The morphology of the colonies differed after 14 days, the same time when the snowflake flocs were observed. Most likely, a mixture of multicellular clusters and single cells was spotted, causing the presence of radial differences in colony colour. A flocculation assay was attempted, but the aggregates could not be deflocculated by adding deflocculation buffer. Deflocculation was also not successful after sonication and heating up to 90 °C during 15 min in the presence of 8 M urea.

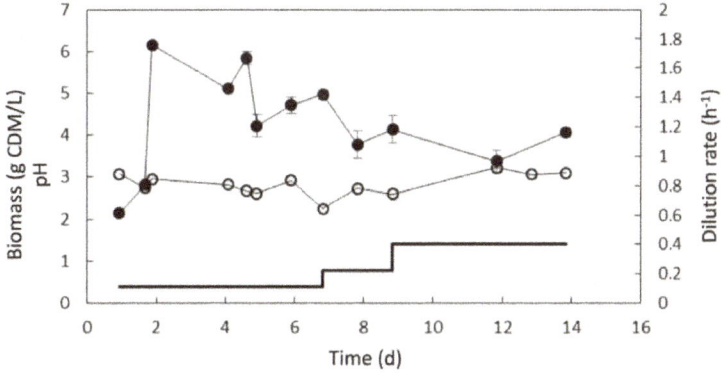

Figure 3. Cultivation of CMBSVM22 in the 3D-printed continuous mini bioreactor. The evolution of the biomass concentration (g·cell dry mass (CDM)/L) (●) of the effluent stream, the pH (○), and the dilution rate (—) during the evolution experiment.

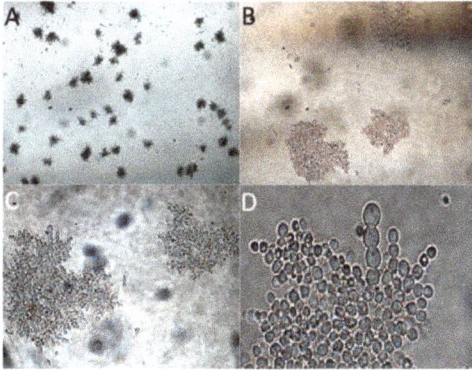

Figure 4. Microscopic observations of snowflake flocs at the end of the fermentation. Objectives: (**A**) 10×; (**B**) 20×; (**C**) zoom-in of 40×; (**D**) 10×.

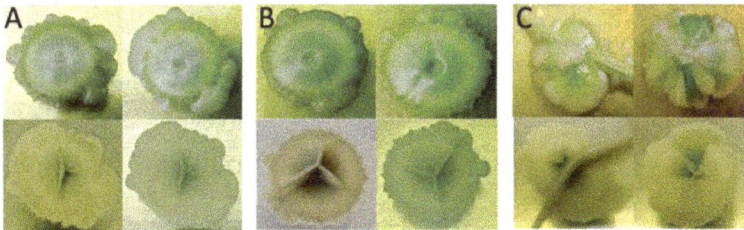

Figure 5. Colonies of CMBSVM22 grown on WL nutrient agar plates. Top (upper panels) and bottom (lower panels) view of two colonies at (**A**) the start of the cultivation; (**B**) after 7 days of cultivation; and (**C**) after 14 days of cultivation.

4. Discussion

A 3D-printed reactor was created, based on an industrial tower fermentor. The reactor is characterized by a high aspect-ratio, which creates a gravity-based selective pressure: single cells have a higher probability to be washed out than cell aggregates, as the denser aggregates tend to sediment faster. Consequently, this reactor type ensures excellent conditions for strain evolution towards a more aggregative phenotype.

We used a commercial physics modelling software package that includes 3D modelling tools. The created 3D model can be exported as an "stl" file that was imported in the print program. The flow of particles such as yeast cells and flocs in the 3D model of the tower fermentor was calculated with the modelling software. In silico simulations confirmed the gravity-based selective pressure in the reactor, and the theoretical retainment advantage of aggregating cells in the reactor was demonstrated.

The employment of 3D-printing for the development of a bioprocess has several advantages. As demonstrated in this contribution, the flexibility of adapting the design allows the design to be optimized quickly. Depending on the selected print resolution, the mini reactor can be printed in a few hours (4–7 h). The used commercial resin is biocompatible, and the printed structure can be autoclaved. We selected transparent resin, which allowed visual observation of the content.

A long-term continuous cultivation experiment was performed to improve the flocculation characteristics of a brewer's ale strain. Cell flocculation is of particular interest to the brewing industry, since it simplifies yeast removal at the end of the fermentation [29]. Still, many industrial brewing strains show no adequate flocculation. ALE is an interesting method to adapt the strain, since genetic manipulation—which is not allowed in the manufacture of products for human consumption—is not involved. At the end of the evolution experiment, yeast cell aggregates were observed with "snowflake" morphology. These multicellular structures did not disaggregate in EDTA buffer nor urea buffer, which indicates that the Flo proteins are not involved in the aggregation mechanism [50,51]. A high temperature of 90 °C could also not destroy the aggregates, which confirms that the aggregation mechanism is not flocculin-based, since these flocs can be disaggregated above 60 °C [50].

A similar morphology of yeast cell clusters was observed when mother/daughter separation was impaired [46]. This phenotype was recently discovered by performing a gravity-based repeated batch evolution experiment with a diploid *S. cerevisiae* Y55 strain where sedimented cells were collected and cultivated further in a next batch culture [44]. It was demonstrated that these multicellular clusters are uniclonal and appeared as "snowflakes". This phenotype was also obtained in an ALE experiment with a haploid *S. cerevisiae* CEN.PK113-7D strain using long-term cultivation in sequential batch reactors [45]. It was shown that a frameshift mutation in *ACE2*—which encodes a transcriptional regulator involved in cell cycle control and mother–daughter cell separation (MDS)—caused the snowflake phenotype. The disruption of the transcription factor Ace2p prevented the splitting of mother and daughter cell [52], and affected the expression of its targets. One of the targets of Ace2p is Cts1p, which is a chitinase that is required for the degradation of the cell septum. Gene expression of *CTS1* was approximately 90% reduced in the evolved strain [45]. Six other genes (*DSE4*, *DSE2*, *SUN4*, *DSE1*, *SCW11*, and *AMN1*) that are regulated by *ACE2* were significantly downregulated in the snowflake yeast [46]. These seven downregulated genes are involved in daughter cell separation, many acting directly to degrade the bud neck septum as MDS enzymes [52–55]. It was shown that *ACE2* knockouts formed cellular clusters [45,53,56,57]. The knockout of *CBK1* (a serine/threonine protein kinase of the RAM signalling network that regulates localisation and activity of Ace2p and the Ssd1p translation repressor) also results in cell clusters [56,58,59]. These results indicate that a lack of the degradation of the septum between the mother and daughter cells could explain the observed multicellular clusters. Whole genome sequencing of the evolved strain that was obtained in the current experiment is necessary to find out if the observed aggregates are the result of (a) mutation(s) in genes involved in MDS, or if mutations in other genes can also result in the snowflake phenotype.

Fermentation **2017**, *3*, 4

5. Conclusions

A continuous mini tower fermentor was designed and constructed using a 3D printer. The printed fermentor was biocompatible for yeast cultivation and could be sterilized by autoclaving. The tower fermentor was suitable to perform experimental evolution experiments where gravity acts as the selective pressure, and allowed yeast cells to evolve from planktonic single cells towards cell aggregates. After 14 days of continuous cultivation of a non-flocculating industrial brewer's yeast strain in the fermentor, yeast flocs with a "snowflake" morphology were observed. Yeast floc characterisation showed that the cell–cell interactions in the cell aggregates were not the result of flocculin interactions, but probably the result of (a) mutation(s) in gene(s) that are involved in the mother/daughter separation process.

Acknowledgments: The Belgian Federal Science Policy Office (Belspo) and the European Space Agency (ESA) PRODEX program supported this work. The Research Council of the Vrije Universiteit Brussel (Belgium) and the University of Ghent (Belgium) are acknowledged to support the Alliance Research Group VUB-UGent NanoMicrobiology (NAMI), and the International Joint Research Group (IJRG) VUB-EPFL BioNanotechnology & NanoMedicine (NANO).

Author Contributions: Ronnie G. Willaert and Andreas Conjaerts conceived and designed the experiments; Andreas Conjaerts performed the experiments; Andreas Conjaerts and Ronnie G. Willaert analysed the data; Ronnie G. Willaert contributed reagents/materials/analysis tools; Ronnie G. Willaert and Andreas Conjaerts wrote the paper.

Conflicts of Interest: The authors declare no conflict of interest.

Appendix A

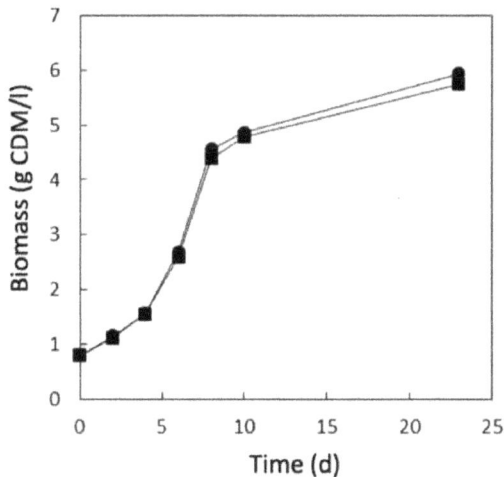

Figure A1. Biocompatibility assay of *S. cerevisiae* BY4742. Evolution of the biomass concentration in the presence (●) and absence (■) of the methacrylic acid ester resin printed material during 1 day of growth.

References

1. Dequin, S. The potential of genetic engineering for improving brewing, wine-making and baking yeasts. *Appl. Microbiol. Biotechnol.* **2001**, *56*, 577–588. [CrossRef] [PubMed]
2. Saerens, S.M.; Duong, C.T.; Nevoigt, E. Genetic improvement of brewer's yeast: Current state, perspectives and limits. *Appl. Microbiol. Biotechnol.* **2010**, *86*, 1195–1212. [CrossRef] [PubMed]

3. Conrad, T.M.; Lewis, N.E.; Palsson, B.Ø. Microbial laboratory evolution in the era of genome-scale science. *Mol. Syst. Biol.* **2011**, *7*, 509. [CrossRef] [PubMed]

4. Portnoy, V.A.; Bezdan, D.; Zengler, K. Adaptive laboratory evolution—Harnessing the power of biology for metabolic engineering. *Curr. Opin. Biotechnol.* **2011**, *22*, 590–594. [CrossRef] [PubMed]

5. Dragosits, M.; Mattanovich, D. Adaptive laboratory evolution—Principles and applications for biotechnology. *Microb. Cell Fact.* **2013**, *12*, 64. [CrossRef] [PubMed]

6. Gibson, B.; Liti, G. *Saccharomyces pastorianus*: Genomic insights inspiring innovation for industry. *Yeast* **2015**, *32*, 17–27. [CrossRef] [PubMed]

7. Blieck, L.; Toye, G.; Dumortier, F.; Verstrepen, K.J.; Delvaux, F.R.; Thevelein, J.M.; van Dijck, P. Isolation and characterization of brewer's yeast variants with improved fermentation performance under high-gravity conditions. *Appl. Environ. Microbiol.* **2007**, *73*, 815–824. [CrossRef] [PubMed]

8. Huuskonen, A.; Markkula, T.; Vidgren, V.; Lima, L.; Mulder, L.; Geurts, W.; Walsh, M.; Londesborough, J. Selection from industrial lager yeast strains of variants with improved fermentation performance in very-high-gravity worts. *Appl. Environ. Microbiol.* **2010**, *76*, 1563–1573. [CrossRef] [PubMed]

9. Yu, Z.; Zhao, H.; Li, H.; Zhang, Q.; Lei, H.; Zhao, M. Selection of *Saccharomyces pastorianus* variants with improved fermentation performance under very high gravity wort conditions. *Biotechnol. Lett.* **2012**, *34*, 365–370. [CrossRef] [PubMed]

10. James, T.C.; Usher, J.; Campbell, S.; Bond, U. Lager yeasts possess dynamic genomes that undergo rearrangements and gene amplification in response to stress. *Curr. Genet.* **2008**, *53*, 139–152. [CrossRef] [PubMed]

11. Ekberg, J.; Rautio, J.; Mattinen, L.; Vidgren, V.; Londesborough, J.; Gibson, B.R. Adaptive evolution of the lager brewing yeast *Saccharomyces pastorianus* for improved growth under hyperosmotic conditions and its influence on fermentation performance. *FEMS Yeast Res.* **2013**, *13*, 335–349. [CrossRef] [PubMed]

12. Strejc, J.; Siříštova, L.; Karabín, M.; Almeida e Silva, J.B.; Brányik, T. Production of alcohol-free beer with elevated amounts of flavouring compounds using lager yeast mutants. *J. Inst. Brew.* **2013**, *119*, 149–155. [CrossRef]

13. Cadière, A.; Ortiz-Julien, A.; Camarasa, C.; Dequin, S. Evolutionary engineered *Saccharomyces cerevisiae* wine yeast strains with increased in vivo flux through the pentose phosphate pathway. *Metab. Eng.* **2011**, *13*, 263–271. [CrossRef] [PubMed]

14. Cadière, A.; Aguera, E.; Caillé, S.; Ortiz-Julien, A.; Dequin, S. Pilot-scale evaluation the enological traits of a novel, aromatic wine yeast strain obtained by adaptive evolution. *Food Microbiol.* **2012**, *32*, 332–337. [CrossRef] [PubMed]

15. Dunham, M.J.; Kerr, E.O.; Miller, A.W.; Payen, C. Chemostat culture for yeast physiology and experimental evolution. *Cold Spring Harb. Protoc.* **2016**. [CrossRef]

16. Miller, A.W.; Befort, C.; Kerr, E.O.; Dunham, M.J. Design and use of multiplexed chemostat arrays. *J. Vis. Exp.* **2013**, *72*, e50262. [CrossRef] [PubMed]

17. Kumar, S.; Wittmann, C.; Heinzle, E. Minibioreactors. *Biotechnol. Lett.* **2004**, *26*, 1–10. [CrossRef] [PubMed]

18. Betts, J.I.; Baganz, F. Miniature bioreactors: Current practices and future opportunities. *Microb. Cell Fact.* **2006**, *5*, 21. [CrossRef] [PubMed]

19. Hortsch, R.; Weuster-Botz, D. Milliliter-scale stirred tank reactors for the cultivation of microorganisms. *Adv. Appl. Microbiol.* **2010**, *73*, 61–82. [PubMed]

20. Tuomi, J.; Paloheimo, K.S.; Vehviläinen, J.; Björkstrand, R.; Salmi, M.; Huotilainen, E.; Kontio, R.; Rouse, S.; Gibson, I.; Mäkitie, A.A. A novel classification and online platform for planning and documentation of medical applications of additive manufacturing. *Surg. Innov.* **2014**, *21*, 553–559. [CrossRef] [PubMed]

21. Mironov, V.; Boland, T.; Trusk, T.; Forgacs, G.; Markwald, R.R. Organ printing: Computer-aided jet-based 3D tissue engineering. *Trends Biotechnol.* **2003**, *21*, 157–161. Erratum in: *Trends Biotechnol.* **2004**, *22*, 265. [CrossRef]

22. Mironov, V.; Kasyanov, V.; Markwald, R.R. Organ printing: From bioprinter to organ biofabrication line. *Curr. Opin. Biotechnol.* **2011**, *22*, 667–673. [CrossRef] [PubMed]

23. Leukers, B.; Gülkan, H.; Irsen, S.H.; Milz, S.; Tille, C.; Schieker, M.; Seitz, H. Hydroxyapatite scaffolds for bone tissue engineering made by 3D printing. *J. Mater. Sci. Mater. Med.* **2005**, *16*, 1121–1124. [CrossRef] [PubMed]

24. Salmi, M. Possibilities of preoperative medical models made by 3D printing or additive manufacturing. *J. Med. Eng.* **2016**, *2016*, 6191526. [CrossRef] [PubMed]
25. Sears, N.A.; Seshadri, D.R.; Dhavalikar, P.S.; Cosgriff-Hernandez, E. A review of three-dimensional printing in tissue engineering. *Tissue Eng. Part B Rev.* **2016**, *22*, 298–310. [CrossRef] [PubMed]
26. Dong, Y.; Fan, S.Q.; Shen, Y.; Yang, J.X.; Yan, P.; Chen, Y.P.; Li, J.; Guo, J.S.; Duan, X.M.; Fang, F.; et al. A novel bio-carrier fabricated using 3D printing technique for wastewater treatment. *Sci. Rep.* **2015**, *5*, 12400. [CrossRef] [PubMed]
27. Gross, B.C.; Erkal, J.L.; Lockwood, S.Y.; Chen, C.; Spence, D.M. Evaluation of 3D printing and its potential impact on biotechnology and the chemical sciences. *Anal. Chem.* **2014**, *86*, 3240–3253. [CrossRef] [PubMed]
28. Melchels, F.P.; Feijen, J.; Grijpma, D.W. A review on stereolithography and its applications in biomedical engineering. *Biomaterials* **2010**, *31*, 6121–6130. [CrossRef] [PubMed]
29. Boulton, C.; Quain, D. Fermentation systems. In *Brewing Yeast and Fermentation*; Blackwell Science Ltd.: Oxford, UK, 2001; pp. 260–372.
30. Klopper, W.J.; Roberts, R.H.; Royston, M.G.; Ault, R.G. Continuous fermentation in a tower fermenter. *Proc. Eur. Brew. Conv.* **1965**, 238–259.
31. Ault, R.G.; Hampton, A.N.; Newton, R.; Roberts, R.H. Biological and biochemical aspects of tower fermentation. *J. Inst. Brew.* **1969**, *75*, 260–277. [CrossRef]
32. Van Mulders, S.E.; Ghequire, M.; Daenen, L.; Verbelen, P.J.; Verstrepen, K.J.; Delvaux, F.R. Flocculation gene variability in industrial brewer's yeast strains. *Appl. Microbiol. Biotechnol.* **2010**, *88*, 1321–1331. [CrossRef] [PubMed]
33. Brachmann, C.B.; Davies, A. Designer deletion strains derived from *Saccharomyces cerevisiae* S288C: A useful set of strains and plasmids for PCR-mediated gene disruption and other applications. *Yeast* **1998**, *14*, 115–132. [CrossRef]
34. Van Mulders, S.E.; Christianen, E.; Saerens, S.M.G.; Daenen, L.; Verbelen, P.J.; Willaert, R.; Verstrepen, K.J.; Delvaux, F.R. Phenotypic diversity of Flo protein family-mediated adhesion in *Saccharomyces cerevisiae*. *FEMS Yeast Res.* **2009**, *9*, 178–190. [CrossRef] [PubMed]
35. Richards, M. The use of giant-colony morphology for the differentiation of brewing yeasts. *J. Inst. Brew.* **1967**, *73*, 162–166. [CrossRef]
36. Pallmann, C.L.; Brown, J.A.; Olineka, T.L.; Cocolin, L.; Mills, D.A.; Bisson, L.F. Use of WL medium to profile native flora fermentations. *Am. J. Enol. Vitic.* **2001**, *52*, 198–203.
37. D'Hautcourt, O.; Smart, K.A. Measurement of brewing yeast flocculation. *J. Am. Soc. Brew. Chem.* **1999**, *57*, 123–128.
38. Den Blanken, J.G. The tower fermenter for lager beer production. *Brew. Gardian* **1974**, 35–39.
39. Seddon, A.W. Continuous tower fermentation—Experiences in establishing large scale commercial production. *Brewer* **1976**, 72–78.
40. Haddad, S.A.; Lindegren, C.C. A method for determining the weight of an individual yeast cell. *Appl. Microbiol.* **1953**, *1*, 153–156. [PubMed]
41. Davis, R.H.; Hunt, T.P. Modeling and measurement of yeast flocculation. *Biotechnol. Prog.* **1986**, *2*, 91–97. [CrossRef] [PubMed]
42. Fontana, A.; Bore, C.; Ghommidh, C.; Guiraud, J.P. Structure and sucrose hydrolysis activity of *Saccharomyces cerevisiae* aggregates. *Biotechnol. Bioeng.* **1992**, *40*, 475–482. [CrossRef] [PubMed]
43. Van Hamersveld, E.H.; Van Der Lans, R.G.; Luyben, K.C. Quantification of brewers' yeast flocculation in a stirred tank: Effect of physical parameters on flocculation. *Biotechnol. Bioeng.* **1997**, *56*, 190–200. [CrossRef]
44. Ratcliff, W.C.; Denison, R.F.; Borrello, M.; Travisano, M. Experimental evolution of multicellularity. *Proc. Natl. Acad. Sci. USA* **2012**, *109*, 1595–1600. [CrossRef] [PubMed]
45. Oud, B.; Guadalupe-Medina, V.; Nijkamp, J.F.; de Ridder, D.; Pronk, J.T.; van Maris, A.J.; Daran, J.M. Genome duplication and mutations in *ACE2* cause multicellular, fast-sedimenting phenotypes in evolved *Saccharomyces cerevisiae*. *Proc. Natl. Acad. Sci USA* **2013**, *110*, E4223–E4231. [CrossRef] [PubMed]
46. Ratcliff, W.C.; Fankhauser, J.D.; Rogers, D.W.; Greig, D.; Travisano, M. Origins of multicellular evolvability in snowflake yeast. *Nat. Commun.* **2015**, *6*, 6102. [CrossRef] [PubMed]
47. Guo, B.; Styles, C.A.; Feng, Q.; Fink, G.R. A *Saccharomyces* gene family involved in invasive growth, cell-cell adhesion, and mating. *Proc. Natl. Acad. Sci. USA* **2000**, *97*, 12158–12163. [CrossRef] [PubMed]

48. Goossens, K.V.; Willaert, R.G. The *N*-terminal domain of the Flo11 protein from Saccharomyces cerevisiae is an adhesin without mannose-binding activity. *FEMS Yeast Res.* **2012**, *12*, 78–87. [CrossRef] [PubMed]

49. Kraushaar, T.; Brückner, S.; Veelders, M.; Rhinow, D.; Schreiner, F.; Birke, R.; Pagenstecher, A.; Mösch, H.U.; Essen, L.O. Interactions by the fungal Flo11 adhesin depend on a fibronectin Type III-like adhesin domain girdled by aromatic bands. *Structure* **2015**, *23*, 1005–1017. [CrossRef] [PubMed]

50. Mill, P.J. The nature of the interactions between flocculent cells in the flocculation of *Saccharomyces cerevisiae*. *J. Gen. Microbiol.* **1964**, *35*, 61–68. [CrossRef] [PubMed]

51. Stratford, M.; Keenan, M.H. Yeast flocculation: Quantification. *Yeast* **1988**, *4*, 107–115. [CrossRef] [PubMed]

52. Weiss, E.L. Mitotic exit and separation of mother and daughter cells. *Genetics* **2012**, *192*, 1165–1202. [CrossRef] [PubMed]

53. King, L.; Butler, G. Ace2p, a regulator of *CTS1* (chitinase) expression, affects pseudohyphal production in *Saccharomyces cerevisiae*. *Curr. Genet.* **1998**, *34*, 183–191. [CrossRef] [PubMed]

54. O'Conallain, C.; Doolin, M.T.; Taggart, C.; Thornton, F.; Butler, G. Regulated nuclear localisation of the yeast transcription factor Ace2p controls expression of chitinase (*CTS1*) in *Saccharomyces cerevisiae*. *Mol. Gen. Genet.* **1999**, *262*, 275–282. [CrossRef] [PubMed]

55. McBride, H.J.; Yu, Y.; Stillman, D.J. Distinct regions of the Swi5 and Ace2 transcription factors are required for specific gene activation. *J. Biol. Chem.* **1999**, *274*, 21029–21036. [CrossRef] [PubMed]

56. Bidlingmaier, S.; Weiss, E.L.; Seidel, C.; Drubin, D.G.; Snyder, M. The Cbk1p pathway is important for polarized cell growth and cell separation in *Saccharomyces cerevisiae*. *Mol. Cell. Biol.* **2001**, *21*, 2449–2462. [CrossRef] [PubMed]

57. Voth, W.P.; Olsen, A.E.; Sbia, M.; Freedman, K.H.; Stillman, D.J. *ACE2*, *CBK1*, and *BUD4* in budding and cell separation. *Eukaryot Cell* **2005**, *4*, 1018–1028. [CrossRef] [PubMed]

58. Versele, M.; Thevelein, J.M. Lre1 affects chitinase expression, trehalose accumulation and heat resistance through inhibition of the Cbk1 protein kinase in *Saccharomyces cerevisiae*. *Mol. Microbiol.* **2001**, *41*, 1311–1326. [CrossRef] [PubMed]

59. Weiss, E.L.; Kurischko, C.; Zhang, C.; Shokat, K.; Drubin, D.G.; Luca, F.C. The *Saccharomyces cerevisiae* Mob2p-Cbk1p kinase complex promotes polarized growth and acts with the mitotic exit network to facilitate daughter cell-specific localization of Ace2p transcription factor. *J. Cell Biol.* **2002**, *158*, 885–900. [CrossRef] [PubMed]

MDPI AG

St. Alban-Anlage 66

4052 Basel, Switzerland

Tel. +41 61 683 77 34

Fax +41 61 302 89 18

http://www.mdpi.com

Fermentation Editorial Office

E-mail: fermentation@mdpi.com

http://www.mdpi.com/journal/fermentation

www.ingramcontent.com/pod-product-compliance
Lightning Source LLC
Chambersburg PA
CBHW051904210326
41597CB00033B/6017